The Improvised State

RGS-IBG Book Series

Published

Learning the City: Knowledge and Translocal Assemblage
Colin McFarlane

Globalizing Responsibility: The Political Rationalities of Ethical Consumption
Clive Barnett, Paul Cloke, Nick Clarke and Alice Malpass

Domesticating Neo-Liberalism: Spaces of Economic Practice and Social Reproduction in Post-Socialist Cities
Alison Stenning, Adrian Smith, Alena Rochovská and Dariusz Świątek

Swept Up Lives? Re-envisioning the Homeless City
Paul Cloke, Jon May and Sarah Johnsen

Aerial Life: Spaces, Mobilities, Affects
Peter Adey

Millionaire Migrants: Trans-Pacific Life Lines
David Ley

State, Science and the Skies: Governmentalities of the British Atmosphere
Mark Whitehead

Complex Locations: Women's Geographical Work in the UK 1850–1970
Avril Maddrell

Value Chain Struggles: Institutions and Governance in the Plantation Districts of South India
Jeff Neilson and Bill Pritchard

Queer Visibilities: Space, Identity and Interaction in Cape Town
Andrew Tucker

Arsenic Pollution: A Global Synthesis
Peter Ravenscroft, Hugh Brammer and Keith Richards

Resistance, Space and Political Identities: The Making of Counter-Global Networks
David Featherstone

Mental Health and Social Space: Towards Inclusionary Geographies?
Hester Parr

Climate and Society in Colonial Mexico: A Study in Vulnerability
Georgina H. Endfield

Geochemical Sediments and Landscapes
Edited by David J. Nash and Sue J. McLaren

Driving Spaces: A Cultural-Historical Geography of England's M1 Motorway
Peter Merriman

Badlands of the Republic: Space, Politics and Urban Policy
Mustafa Dikeç

Geomorphology of Upland Peat: Erosion, Form and Landscape Change
Martin Evans and Jeff Warburton

Spaces of Colonialism: Delhi's Urban Governmentalities
Stephen Legg

People/States/Territories
Rhys Jones

Publics and the City
Kurt Iveson

After the Three Italies: Wealth, Inequality and Industrial Change
Mick Dunford and Lidia Greco

Putting Workfare in Place
Peter Sunley, Ron Martin and Corinne Nativel

Domicile and Diaspora
Alison Blunt

Geographies and Moralities
Edited by Roger Lee and David M. Smith

Military Geographies
Rachel Woodward

A New Deal for Transport?
Edited by Iain Docherty and Jon Shaw

Geographies of British Modernity
Edited by David Gilbert, David Matless and Brian Short

Lost Geographies of Power
John Allen

Globalizing South China
Carolyn L. Cartier

Geomorphological Processes and Landscape Change: Britain in the Last 1000 Years
Edited by David L. Higgitt and E. Mark Lee

The Improvised State: Sovereignty, Performance and Agency in Dayton Bosnia
Alex Jeffrey

Forthcoming

Spatial Politics: Essays for Doreen Massey
Edited by David Featherstone and Joe Painter

In the Nature of Landscape: Cultural Geography on the Norfolk Broads
David Matless

Working Memories: Gender and Migration in Post-war Britain
Linda McDowell

Fashioning Globalisation: New Zealand Design, Working Women and the 'New Economy'
Maureen Molloy and Wendy Larner

Dunes: Dynamics, Morphology and Geological History
Andrew Warren

Scalar Politics of Food in Cuba
Marisa Wilson

The Improvised State

Sovereignty, Performance and Agency in Dayton Bosnia

Alex Jeffrey

WILEY-BLACKWELL

A John Wiley & Sons, Ltd., Publication

This edition first published 2013
© 2013 John Wiley & Sons, Ltd.

Wiley-Blackwell is an imprint of John Wiley & Sons, formed by the merger of Wiley's global Scientific, Technical and Medical business with Blackwell Publishing.

Registered Office
John Wiley & Sons, Ltd, The Atrium, Southern Gate, Chichester, West Sussex, PO19 8SQ, UK

Editorial Offices
350 Main Street, Malden, MA 02148-5020, USA
9600 Garsington Road, Oxford, OX4 2DQ, UK
The Atrium, Southern Gate, Chichester, West Sussex, PO19 8SQ, UK

For details of our global editorial offices, for customer services, and for information about how to apply for permission to reuse the copyright material in this book please see our website at www.wiley.com/wiley-blackwell.

The right of Alex Jeffrey to be identified as the author of this work has been asserted in accordance with the UK Copyright, Designs and Patents Act 1988.

Library of Congress Cataloging-in-Publication Data

Jeffrey, Alexander Sam.
The improvised state : sovereignty, performance and agency in Dayton Bosnia / Alex Jeffrey.
    p. cm.
  Includes bibliographical references and index.
  ISBN 978-1-4443-3699-3 (cloth) – ISBN 978-1-4443-3700-6 (pbk.)
 1. Bosnia and Hercegovina–Politics and government–1992– 2. Sovereignty. 3. Geopolitics–Bosnia and Hercegovina. 4. Dayton Peace Accords (1995)
  JN2203.A58J44 2012
  320.94974–dc23
                                                                      2012015743

A catalogue record for this book is available from the British Library.

Cover image © Alex Jeffrey
Cover design by Workhaus

Set in 10/12pt Plantin by SPi Publisher Services, Pondicherry, India

Printed in Malaysia by Ho Printing (M) Sdn Bhd

1 2013

Dedicated to the memory of
Ellie Maxwell (1977–2009)

Contents

List of Figures

Series Editors' Preface

The RGS-IBG Book Series only publishes work of the highest international standing. Its emphasis is on distinctive new developments in human and physical geography, although it is also open to contributions from cognate disciplines whose interests overlap with those of geographers. The Series places strong emphasis on theoretically informed and empirically strong texts. Reflecting the vibrant and diverse theoretical and empirical agendas that characterize the contemporary discipline, contributions are expected to inform, challenge and stimulate the reader. Overall, the RGS-IBG Book Series seeks to promote scholarly publications that leave an intellectual mark and change the way readers think about particular issues, methods or theories.

For details on how to submit a proposal please visit:
www.rgsbookseries.com

Neil Coe
University of Manchester, UK

Joanna Bullard
Loughborough University, UK

RGS-IBG Book Series Editors

Acknowledgements

The arguments in this book have been developed over a decade of researching international intervention in Bosnia and Herzegovina. Part of the writing emerges from doctoral work conducted at the Geography Department at Durham University and funded by the Economic and Social Research Council (ESRC award number R42200134266). I am extremely grateful to my three supervisors, Joe Painter, Emma Mawdsley and Luiza Bialasiewicz, for their support and advice over this project and beyond. The arguments were refined through an ESRC-funded post-doctoral fellowship (ESRC award number PTA-026-27-0576), also at Durham University, and I would like to thank Ash Amin for his guidance and feedback through this process. Finally, the book proposal and the majority of the writing were completed while I was lecturing at Newcastle University's School of Geography, Politics and Sociology. I would like to thank my colleagues over this period, in particular Nick Megoran, Matthew Rech, Alison Williams, Fiona McConnell, Rachel Woodward, Nina Laurie, Andy Gillespie, Martin Coward and Stuart Dawley. I am especially grateful to Raksha Pande who provided assistance with synthesizing literatures on improvisation and the state. Beyond these institutional contexts I am grateful to a range of people for advice and support over the writing period, in particular Colin McFarlane, Alex Vasudevan, Carl Dahlman, Dan Swanton, Peter Thomas, Briony Jones, Merje Kuus, Stuart Elden, David Campbell, Klaus Dodds and Lynn Staeheli.

The process of writing the book has been made considerably more straightforward by the expertise and support of those on the RGS-IBG Book Series editorial team; in particular I would like to thank Kevin Ward and Neil Coe for their extremely supportive and constructive editorial skills. I would also like to thank Jacqueline Scott and Isobel Bainton at Wiley-Blackwell for their advice and encouragement. The proposal and manuscript were strengthened through the feedback and suggestions of the anonymous referees; I am grateful for their assistance. I am also grateful to the publishers for the permission to draw on material from previously published material in three of the chapters of this book. Chapter Four contains material previously published in *Political Geography* 25(2) 203–227 (Elsevier, Philadelphia); Chapter Five contains discussions previously published in *Development and Change* 38(2) 251–274 (Wiley-Blackwell, Oxford), and a version of Chapter Seven appeared in *Environment and Planning D: Society and Space* 26(3) 428–443 (Pion, London).

There are a large number of people to thank who assisted with the empirical research for this book. In Brčko District I would like to thank Goran Mihailović, Saška Haramina, Mirella, Davor and Aleksa Ceran, Leila Jaserević, Gordana Varcaković, Kristina Varcaković, Elenora Emkić, Jack Richold, Catharina de Lange, Slawomir Klimkiewicz and Mary Lynch. In Sarajevo I would like to thank Zlatan Music, Asim Mujkić, Damir Arsenijević, Refik Hodzić and Selma Hadzić. I would also like to thank Matt Bolton for his support and friendship since we met in Brčko in 2002. Many of the arguments in the book began as discussions with Matt, and I have benefited greatly from his wisdom.

Over the final months of the book writing I am indebted to the support and friendship of Michaelina Jakala, as research assistant on a two-year study of the public outreach strategies of the State Court of Bosnia and Herzegovina (ESRC award number RES-061-25-0479). I am grateful for her insightful reading of the manuscript and her wise advice. I am also grateful to her husband Martin and daughter Nia for being such great hosts in Sarajevo.

I would like to thank Craig Jeffrey, Jane Dyson, Ewan Jeffrey and my parents for all the help over the years, in particular Craig's assistance in thinking through improvisation. The biggest thanks go to Laura Jeffrey. It certainly wouldn't have been possible to complete this project without her support and help; I am forever grateful for all the interest, the careful readings, the ideas and the encouragement. A final big thanks and hug to Rufus and Clemence for making the time between writing so much fun.

This book is dedicated to Ellie Maxwell, an inspirational friend and colleague who worked tirelessly over her lifetime for causes in Bosnia and Herzegovina, particularly in the town of Brčko. Ellie and I met as fellow students at Edinburgh University, and it was through working for Firefly Youth Project in Brčko in 1999–2000 that I first became interested in the question of state building in Bosnia. Ellie is greatly missed, and all who knew her continue to be inspired by her insights, intelligence and compassion.

Abbreviations

ARBiH	*Armija Republike Bosne i Hercegovine* (Army of the Bosnian Republic)
BiH	Bosnia and Herzegovina
CBiH	The Court of Bosnia and Herzegovina
CCI	*Centri civilnih inicijativa* (Centre for Civil Initiative)
CSN	Court Support Network
DMT	District Management Team (in Brčko District)
DP	Displaced Person
EU	European Union
FRY	Federal Republic of Yugoslavia
GFAP	General Framework Agreement for Peace (Dayton Peace Accords)
GTZ	*Gesellschaft für Technische Zusammenarbeit* (German Technical Cooperation Agency)
ICG	International Crisis Group
ICTY	International Criminal Tribunal for the former Yugoslavia
IEBL	Inter-Entity Boundary Line
I-For	Implementation Force
IHC	International Housing Commission
IPTF	International Police Task Force
IRC	International Rescue Committee
JNA	*Jugoslovenska narodna armija* (Yugoslav People's Army)
MZ	*Mjesna zajednica* (local community association)
NATO	North Atlantic Treaty Organization
NGO	Non-Governmental Organization
OHR	Office of the High Representative
OSCE	Organization for Security and Co-operation in Europe
OZNa	*Odeljenje za zastitu narodna* (Department for the People's Defence)
PIC	Peace Implementation Council
RRTF	Return and Reconstruction Task Force

RS	*Republika Srpska* (sub-division of Bosnia and Herzegovina)
SDA	*Stranka demokratska akcije* (Party for Democratic Action)
SDP	*Socijaldemokratska partija Bosne i Hercegovine* (Social Democratic Party)
SDS	*Srpska demokratska stranka* (Serb Democratic Party)
S-For	Stabilization Force
SIDA	Swedish International Development Corporation Agency
SNSD	*Savez nezavisnih socijaldemokrata* (Alliance of Independent Social Democrats)
UN	United Nations
UNDP	United Nations Development Programme
UNHCR	United Nations High Commission for Refugees
USAID	United States Agency for International Development
USDA	United States Department for Agriculture
VOPP	Vance Owen Peace Plan
VRS	*Vojska Republike Srpske* (Army of Republika Srpska)
WCC	War Crimes Chamber
ZOS	Zone of Separation

Chapter One

Introduction

Figure 1   Palestinian chair at the United Nations
Source: © Stan Honda/AFP/Getty Images

New York, 23 September 2011. The head of the Palestinian Authority, Mahmoud Abbas, is seeking a vote at the United Nations on an application for Palestinian admission to the UN as a member state. In the build-up to the request for the vote, Palestinian activists have produced a chair as a symbol of the desire for a Palestinian seat at the General Assembly of the United Nations. In the preceding weeks the chair has toured the Middle

The Improvised State: Sovereignty, Performance and Agency in Dayton Bosnia, First Edition. Alex Jeffrey.
© 2013 John Wiley & Sons, Ltd. Published 2013 by John Wiley & Sons, Ltd.

East and Europe, before taking pride of place at news conferences in New York in the lead-up to the vote. The symbolism is easy to grasp: the chair is covered in blue velour, marked with the UN olive branches encircling a symbol of another seat, on which the Palestinian Authority's flag is imprinted. Underneath these images are sewn the words 'Palestine's Right: A full membership in the United Nations'. But underpinning this stark imagery are two more subtle assumptions: the first, that desire for Palestinian statehood could be fulfilled through the recognition granted by UN membership. Membership would serve as a symbol of statehood, despite not necessarily changing the forms of authority or territorial control in the West Bank and Gaza. Indeed, youth activists in Ramallah in the West Bank were keen to distinguish between the 'emotional' nature of international recognition and the unchanging 'practical' everyday experience of militarized check points: settlement construction and inhibited freedom of movement (see BBC, 2011). But the second assumption is reflected in the symbolism of the seat itself. The claim to Palestinian statehood is not made solely in a speech to the General Assembly of the UN, but is rather symbolized through the creation of the seat. The act of producing the seat, and its tour through Europe and the Middle East, provide a chance to perform statehood, to ground the legitimacy and effect of the claim through repeated enactments of the securing of a UN seat. In this sense, performance is at the heart of attempts to convey state legitimacy. The design of the chair draws on audience expectations of a 'real' UN seat, primed as they would be to recognize the appropriate colours and symbolism for UN furniture.

States are improvised. Their legitimacy and ability to lay claim to rule rely on a capacity to perform their power. Such performances of the state are often spectacular: the pageantry of ambassadorial relations, the ceremony of opening parliament, the celebration of a military victory (see Bodnar, 1993; Marston, 1989; McConnell *et al.*, forthcoming; Navaro-Yashin, 2002). But more often they are prosaic: the modes of address and comportment at international borders, the use of headed paper to claim a missing tax return, the statutory warning advice on a bottle of wine (see Painter, 2006; Raento, 2006). As the claims to statehood of the Palestinian Authority attest, performances of the state are often more explicit where changes are desired in the existing inter-state system, where a particular political authority is seeking to assert or solidify a specific claim to the state. But alongside the use of performance lies a secondary part of this opening story: performances are structured by available resources. At first sight the creation of the seat appears to represent Palestinian subjection, where those excluded from formal state structures improvise their own version to illustrate the asymmetry of power relations (Scott, 1985). But the move to create a seat also illustrates the prevailing resources available to those seeking to contest the existing state system. In the case of the Palestinian Authority they drew on conventions of colour and imagery to lend legitimacy to the UN seat as a form of state symbolism. While they could not claim an 'official' seat, the production of an 'unofficial seat' demonstrates the aspiration of state recognition. The focus on the UN underlines a wider public expectation of this institution as an arbiter of state legitimacy.

This book argues for an understanding of states as improvisations, where improvisation is conceived as a process that combines performance and resourcefulness. In order to make this argument the book explores the experience of state building in Bosnia and Herzegovina (or BiH[1]). This approach contributes to three areas of existing scholarship. First it develops recent work in the social sciences that has explored the state as an idea or process rather than a stable administrative entity (Abrams, 2006 [1988]; Jones, 2007; Mitchell, 1999; Painter, 2006; Trouillot, 2001). This work has orientated attention on the forms of social and cultural effects produced by the state, arguing that the state should be understood as a human accomplishment rather than the static backdrop to political life (Radcliffe, 2001). Building on this approach this book highlights the forms of agency through which state improvisations are performed, exploring how competing understandings of the state may coexist in everyday life.

Second, this argument contributes to understandings of the state in BiH. Rather than lamenting the 'failure' or 'weakness' of the state, it centres on the forms of practices produced by intensive international intervention since the signing of the Dayton General Framework Agreement for Peace (GFAP) in December 1995. David Campbell (1998a, 1999) provides a nuanced set of illustrations of the territorial, social and democratic consequences of the connection between identity and territory forged at Dayton. This book extends Campbell's analysis by exploring the ongoing political and spatial consequences of the GFAP. Analysis of the practices of the state illustrates the multiple institutions that have been enrolled in performing the state in BiH since 1995, including international agencies, domestic politicians, international non-governmental organizations (NGOs) and local civil society organizations. Rather than making a simple distinction between the power of international elites and the subordination of local political actors, this approach allows the analysis to explore the considerable entanglement of these groups. The theoretical framework of improvisation illuminates the multiple competing claims to state sovereignty that circulate in contemporary BiH. This approach contributes to recent work that has examined the geographically uneven nature of state effects following intervention in BiH, work that has explored refugee return (Toal and Dahlman, 2011), the state in everyday life (Bougarel et al., 2007) and forms of criminal justice reform (Aitchison, 2011).

Third, by exploring the nature and consequences of international intervention in BiH this argument contributes to emerging work studying the production of geopolitical knowledge. Recent scholarship in the fields of critical and feminist geopolitics has looked beyond traditional preoccupations with textual analysis of policy statements to explore the forms of practices and materials that produce geopolitical knowledge. From studies of boundary disputes in Central Asia (Megoran, 2006) through to the circulation of knowledge in the bureaucracy of the European Commission (Kuus, 2011), scholars are looking towards geopolitics as a form of social practice. Understanding the state as an improvisation encourages a reflection on such everyday and banal practices that shape popular understandings of geopolitics. Consequently this study draws on four periods of residential fieldwork, comprising methodologies of extended interviews and participant observation to illustrate state improvisation in BiH and document subsequent forms of political subjectivity.

Examining BiH as an improvised state thus contributes to state theory, understandings of intervention in BiH and methodologies of critical geopolitics. The following sections provide greater detail concerning how the book contributes to these three areas of scholarship, before a concluding section summarizes the book's chapter structure. What is shared across these areas of inquiry is a focus on the forms of political agency established through attempts to create and present a coherent BiH state. This is not an attempt to establish a new overarching theory of the state, or to resolve debates concerning the relative influence of structure or agency in understanding state power. Rather I am seeking to present the utility of improvisation as a means through which to illuminate the contingent and plural nature of state claims in BiH.

1.1   States, Performance and Improvisation

The state is politically and intellectually seductive. At its most basic level, the state is a mechanism for fixing political power to geographical space through establishment of sovereignty over territory. Politically, this is attractive as it provides a form of order that acts 'as if' different state regimes are comparable across both time and space. Intellectually, this has allowed scholars of international relations to theorize political relationships through this pre-given geographical framework. The state becomes the lens through which global political contestation may be analysed and understood. But the state seduces in other ways. It also conjures a notion of distinct and concrete administrative entities that act as the centre of political decision making within a given territory. This is most commonly expressed in the distinction between state and society. Separating state and society allows policy makers to express statecraft as distinct from the messy context of the society within which it is embedded. Instead it can be presented as a form of political logic that is driven by the state's privileged position 'above' the society it serves (Abrams, 2006 [1988]; Ferguson and Gupta, 2002; Sharma and Gupta, 2006). This form of state reasoning leads to the production of intricate technologies to rule a given society, through statistics, cartography and infrastructure. Intellectually this has led to a wealth of studies attempting to understand the interplay between state and society, work that has sought to theorize how one shapes the other (see Painter, 2003).

What stands out in these imaginaries of the state is the act of seduction. The state is not a natural expression of political power; it is a human achievement of control. This achievement is reflected in prevailing definitions of state sovereignty. Take the most regularly cited: Weber's definition of the state as a 'human community that (successfully) claims the monopoly of legitimate use of physical force within a given territory' (Weber, 1958: 78). The human community, physical force and territory are all central components, but the achievement of statehood is granted through the more elusive notion of the conferment of legitimacy (see Bratsis, 2006: 14). The first act of seeking to understand this seductive power of the state is to de-naturalize it, to sever it from an image of pre-existing or inevitable political territorializations. Our assumptions about the state are themselves a reflection of the power of this mental category to shape our thinking and orientate the design of research (Bilgin and Morton, 2002; Jeffrey, 2009a). Over the last century scholars have sought to examine this act of seduction by turning attention to the forms of practices and processes that reproduce the idea of the state. It has been established within Marxist and, more recently, post-structural studies of the state that we need to look beyond the representation of coherence to uncover the subjective processes whereby the idea of the state is conveyed as a stable truth (see Poulantzas, 1978; Weber, 1998).

The book explores the improvised state through the example of international attempts to establish a state of BiH since 1995. I argue that since the signing of GFAP the BiH state has been improvised among a variety of actors operating across a range of spatial scales. The focus on improvisation reflects a wider adoption of this term across the social sciences and humanities, including musical performance (Berliner, 1994), business administration (Baker et al., 2003), education (C. Jeffrey et al., 2008) and constitutional reform (Garvey, 1971). As this work attests, the lens of improvisation highlights the 'doing' of social practice as it is worked through in everyday life. In order to illustrate the practice of the Bosnian state in the post-conflict era, I break down the lens of improvisation into its two constituent parts: *performance* and *resourcefulness*. Taking these two facets in turn, this argument contributes to recent scholarship within the social sciences that has explored the operation of institutions, individuals and identities as 'performed'. The language of performance influences the theoretical context of the book in two ways.

First, *The Improvised State* draws on anti-foundationalist feminist theory that has sought to undermine the stability of essential identities. Following Butler (1997), scholars have argued that gender is not a status but should more readily be understood as a set of performances that reify particular prevailing understandings and hierarchies of gendered identity. This anti-foundationalist lens will provide a framework through which the reproduction of particular categories and assumptions may be understood in post-Dayton BiH. Performances of the 'international community' and 'nationalist politics' are explored as practices rather than expressions of essential identities. It will be shown that acting and speaking about BiH does not simply report reality but actively constitutes and reproduces political categories and territorializations.

Second, the language of performance has been adopted in strands of cultural geography which have identified the inadequacy of textual representation in conveying inter-subjective feelings. This scholarship has argued that performance highlights the embodied and enacted nature of social and political life which in many senses defies the closure of language and text. This has led some scholars to advocate non-representational theory that foregrounds affective responses to place and time (Anderson, 2006; Thrift, 2003). This work highlights performance as a dynamic set of processes conveyed through an assemblage of materials, apparatuses and milieus (Allen and Cochrane, 2007; Featherstone, 2008; McFarlane, 2009). Such debates concerning representation enrich understanding of international intervention by exploring the affective responses to attempts to establish new state practices, from despair at eviction from temporary accommodation to shared experiences of solidarity and hope at political rallies. The engagement with the concept of performance allows a more nuanced understanding of the social context in which statecraft is enacted and identifies spaces of resistance to dominant narratives of state building.

Understanding the state as a set of performances sheds light on the grounding of sovereignty claims in social and cultural practice, but it says rather less on the rationality that informs the selection of individual performances or how performances are contested and reshaped in everyday life. In order to explore these facets of state practice the book develops a second aspect of improvisation: resourcefulness. This aspect of improvisation is indebted to the structural anthropology of Claude Lévi-Strauss (1972), who through his study of belief systems coined the term *bricolage* to intimate the way in which non-Western societies make sense of the world through 'making do' with available social categories and symbols (see also Hebdige, 1979). This approach is rooted in a syntactic understanding of social forms as related through grammar relations, where society fits together 'like words in a sentence, to form a meaningful whole' (Garvey, 1971: 11). This structural approach has been criticized in recent years for underplaying individual agency and failing to account for forms of dominance and exploitation (Werbner, 1986). Acknowledging these criticisms and in order to adequately theorize the agency inherent in resourcefulness, I draw on the work of Pierre Bourdieu (1984, 1989) and in particular his economic metaphor of capital. This conceptualization of capital shares little with Marx's purely economic understanding of the term, and is instead used to illuminate scarcity across social, cultural and symbolic arenas (C. Jeffrey, 2001: 220;

Painter, 2000). For example, in contrast to developmental understandings of social capital as a form of 'societal glue', Bourdieu's work emphasizes the value placed on 'social connections' (Calhoun, 1993: 70) or 'group membership' (Bourdieu, 1987: 4). Similarly, Bourdieu recognized the accumulation of cultural capital in various forms of credentials, in particular in the arena of education (C. Jeffrey *et al.*, 2008). The concepts of social and cultural capital provide a language through which to explore the everyday performances and dispositions that reproduce class advantage. Applying this approach to the performance of the state enables the development of a framework through which certain understandings of sovereignty and space are imbued with value. Adopting a lens of improvisation unsettles the concept of the state as a stable backdrop to political life, and instead analysis shifts to the forms of practice, materials and imaginaries that convey particular understandings of the state.

1.2 Towards a Political Anthropology of the Bosnian State

The BiH state cannot be understood in isolation from the wider rise and fall of Yugoslavia. Neither can this narrative of consolidation and fragmentation be divorced from the prevailing geopolitical interests of other states and powerful agencies. This is not to argue for a form of historical determinism, nor to indulge in the misconception that current BiH politics necessarily requires an understanding of medieval enmities and allegiances. As discussed in Chapter Three, the scholar of BiH needs to assess competing historical claims carefully, grounded as they are in different conceptualizations of just outcomes in the present day. Rather, in order to understand BiH state building it is necessary to examine key moments in its history and in doing so explore how the state of Yugoslavia was made and unmade through deliberate actions that prioritized different understandings of group membership, political authority and territorial claims.

The first state of Yugoslavia, the Kingdom of Serbs, Croats and Slovenes, was established in 1918 through a union of Serbia and Montenegro with the South Slav lands of the former Austro-Hungarian Empire, principally the territories of Slovenia, Croatia, Bosnia and Montenegro (see Hoare, 2010). The unity of this state was undone by the violence of the Second World War, where loyalties fragmented between Croat Ustaše forces, supported by Nazi Germany, Serb-nationalist Četnik groups, loyal to the exiled King Aleksander, and Partisan forces seeking to retain a unified Southern Slav state. Over the course of the conflict Allied support transferred from Serb-nationalist to Partisan forces, in part contributing to the inauguration of their leader, Josep Broz 'Tito', as the President of Yugoslavia from 1947 to 1980. This period of rule was one of intense Yugoslav state consolidation, under the banner of '*bratsvo i jedinstvo*' ('brotherhood and unity'), where political elites sought to relegate group differences, such as those down 'national' or 'ethnic' lines, in order to promote civic solidarity to the state. These initiatives reflected the significance of centrifugal forces that sought to promote different visions of the state, often drawing on the borders of the six republics that comprised Yugoslavia (Slovenia, Croatia, Serbia, BiH, Montenegro and Macedonia) or in terms of unifying a single 'national' or 'ethnic' group within a continuous territory.

The death of Tito in 1980 marked the start of a decade of decline for the Yugoslav state; as economic disparities between the republics grew (though these problems had a longer lineage), nationalist movements began to emerge from the fragmentation of the single-party *Savez komunista Jugoslavije* (League of Communists of Yugoslavia). In the later 1980s key figures such as Slobodan Milošević in Serbia and Franjo Tudjman in Croatia began to speak of the significance of unifying 'their' ethnic group and political territory; in short, cultivating a political discourse of ethnically aligned states. Considering the extensive intermingling of people now aligning themselves with different 'ethnic' groups, the outcome of such political rubrics would always require massive population movements. For example, the 1991 Yugoslav census indicates that there was a population of around 245 800 people in the Krajina region of the Republic of Croatia identifying themselves as Serb, and 760 852 people resident in the Republic of BiH who considered themselves Croat.[2] These demographic realities added complexity to the supposed moral clarity of calls by nationalist politicians for 'self-determination', as such outcomes would produce challenges for newly created minority populations within successor states. This was illustrated in the case of the declarations of independence by Croatia and (to a lesser extent) Slovenia in 1991, where the *Jugoslovenska narodna armija* (Yugoslav People's Army or JNA) was deployed by Slobodan Milošević ostensibly to protect the unity of the Yugoslav state, though perhaps most significantly to defend Serb populations in the Krajina (see Gow, 2003). In turn Tudjman presented the right of Croatian independence in the language of self-determination following a referendum on 19 May 1991. In order to arbitrate on these competing claims to independence, the then European Economic Community established the Arbitration Commission of the Peace Conference on the former Yugoslavia (usually known as the 'Badinter Commission' after its first President, Robert Badinter). Placing particular emphasis on the legal principle of *Uti possidetis* ('as you possess'), across a number of judgments the Badinter Commission sought to preserve the existing borders of the Yugoslav republics, as opposed to granting self-determination to those minorities living within these territories (see Pomerance, 1998).

This tension between the recognition of existing borders and the rights of minority groups became more pronounced in the case of BiH. Following the successful independence claims of Slovenia, Croatia and Macedonia, by the end of 1991 Yugoslavia comprised only BiH, Serbia and Montenegro.

Within BiH the population was demographically mixed between those identifying as Bosnian Muslim, or Bosniak (43.5% of the population), those identifying themselves as Serb (31.2%), those identifying as Croats (17.4%) and those identifying themselves as Yugoslav (5.5%).[3] In February 1992 Alija Izetbegović, then President of the Bosnian Republic and one of the leading figures in the Bosniak-orientated *Stranka demokratske akcije* (Party for Democratic Action, or SDA), called a referendum for independence. The majority of those loyal to Serb causes boycotted the referendum, seeing the vote as both unconstitutional and leading to minority status in a BiH state. Despite this resistance the turnout for the vote was 63%, with 99.4% of voters preferring independence (Bethlehem and Weller, 1997: xxxiv). With this strong democratic mandate Izetbegović claimed independence on 3 March 1992; by 7 April this had been recognized by the European Community and the United States.

The claims to independence and its recognition led to a series of military skirmishes in Sarajevo and the northern Bosnian town of Bosanski Brod (see Bethlehem and Weller, 1997). The violence spread over 1992 as military forces loyal to Serb causes, in particular the newly formed *Vojska Republike Srpske* (Army of Republika Srpska, or VRS) and the remains of the JNA, sought to create an 'ethnically pure' Serb territory by expelling or murdering non-Serb populations (Silber and Little, 1996; Toal and Dahlman, 2011). Of course, such 'ethnic' interpretations of the violence reproduced the categories espoused by perpetrating individuals and groups. However, as documented in Chapters Three and Four, a consensus soon emerged between BiH politicians and intervening agencies that the ethnic matrix was the dominant framework for understanding the violence in BiH, whether or not this reflected the everyday experience of BiH citizens during or prior to the violence (see Bringa, 1995). While figures are disputed, a 2007 report conducted by the Research and Documentation Center Sarajevo and funded by the Norwegian government estimates that over the following three-and-a-half years of conflict 97 207 people lost their lives, around 65% of whom were registered as Bosniak (see Research and Documentation Center Sarajevo, 2007).

The violence in BiH between 1992 and 1995 has prompted considerable scholarly reflection on the causes and consequences of the conflict (Bennett, 1995; Silber and Little, 1996; Sokolovic, 2005; Udovički and Ridgeway, 2000). One of the shared conclusions of this work relates to the danger of attempting to identify a singular cause for the conflict, and in particular challenging the simplistic political refrain that the war was a consequence of 'ancient hatreds'. The quest for a singular explanation to the conflict in BiH requires a violent simplification of complex social, political and economic realities (see Bose, 2002; Campbell, 1998a). As Bougarel *et al.* (2007: 11) note, interpretations of the violence have often been seduced by imaginaries of ethnic social cleavages that ignore the economic, criminal and wider political networks and affiliations that motivated the conflict (see also Andreas, 2008; Bojičić, 1996).

But beyond attempts to explain the causes of the violence, there is an allied debate concerning the geopolitics of the conflict itself. One of the key areas of disagreement among scholars examining the conflict in BiH between 1992 and 1995 is whether this violence should be described as a 'civil war' (comprising antagonists emerging from within the borders of a single state) or a 'foreign aggression' with warring parties crossing state boundaries (see Woodward, 1995, 1996). This argument rests on questions of state recognition, specifically whether the Bosnian declaration of independence in April 1995 marked the creation of an independent sovereign state. If it did, the presence of the JNA is evidence of external aggression orchestrated from the remains of Yugoslavia (then Serbia and Montenegro). If BiH was still part of Yugoslavia, as groups loyal to Serb causes have suggested, it is perhaps easier to present the violence as an 'internal' matter.[4] This geographical debate has two security dimensions. First, defining the state also defines the minority group: those identifying as Serbs would be a minority in the newly independent BiH, while those identifying as Bosniaks feared minority status in the remnants of Yugoslavia. With weak guarantees for minority rights in both states, the nature of the state shaped perceptions of group security. Second, if the conflict is presented as an external aggression there is a moral imperative among UN Security Council members to intervene to halt the violence, as occurred in the case of Iraq's invasion of Kuwait in 1990. This debate illustrates the significance of questions of recognition and legitimacy to understandings of the state and security. States do not simply exist, but are made through acts intended to convince audiences of the legitimacy of particular sovereign arrangements.

The desire to present certain sovereign arrangements in the former Yugoslavia as legitimate has been the focus of academic debate over the past fifteen years. This work has explored a number of aspects of state sovereignty in South East Europe, from the rise of the Yugoslav state following anti-Imperial struggles of the nineteenth and early twentieth centuries (Jambrek, 1975; Udovički and Ridgeway, 2000), through ethno-national projects that sought to link statehood with ethnic identity (Gow, 2003; Judah, 2000), to international attempts to establish BiH following the signing of the GFAP (Bose, 2002; Chandler, 1999). Much of this work has explored the nature of state sovereignty from a 'top-down' perspective, drawing on historical accounts, largely from official archives, examined through methods of discourse and content analysis. This attachment to the primacy of the state reflects what Bilgin and Morton (2002) term the 'embedded statism' of political analysis, where twentieth-century scholarship and politics have demonstrated an enduring attachment to states as territorial 'containers of power' within geopolitical discourse (Taylor, 1994).

The analyst of sovereignty in BiH therefore faces a challenge: to acknowledge the significance of new state arrangements without reducing political life to the activities of the state. One of the theoretical and empirical responses to this challenge has been to examine the localized expression of the state in social practices and institutions. Drawing on perspectives from anthropology and political geography, this work has explored the practices, materials and subjectivities through which ideas of the state and society are conveyed in BiH (see Bougarel *et al.*, 2007). In doing so this work has advanced a critical geopolitics of the state, where the assertion of state territorialities illustrates particular power relations rather than revealing essential truths (Jeffrey, 2007; Ó Tuathail, 1996). Mirroring the discussions of state theory conducted above, this work has begun to explore the techniques through which certain understandings of the Bosnian state are made credible and normal, while others are cast out as criminal or abnormal.

This approach is exemplified in Toal and Dahlman's (2011) exploration of the extent and characteristics of refugee return in BiH after the GFAP. Rather than emphasizing classical geopolitical concerns of Great Power relations, where the nature of the Bosnian state may be derived from imaginaries of global power dynamics, the authors illustrate a more localized and plural account of the state:

> In conditions of crisis, hierarchies of power come undone and alternative connectivities and networks – those created by diasporas, arms smugglers, media networks, cultural activists and Secret Service agents – emerge as competitors and potential replacements. Geopolitics in such conditions is less a determining location or a stable hierarchy than an entrepreneurial field for the creation of networks that subvert and undermine traditional verticals of power and governance. (Toal and Dahlman, 2011: 10)

This analysis draws attention to the challenge of reasserting state authority in conditions of multiple competing claims to power. One of the tensions in this process in BiH has been the attempt to establish a new state architecture while simultaneously presenting imposed forms of rule as democratically legitimate. As the political system in BiH stagnated after the signing of the GFAP, and nationalist political parties continued to strengthen their political support, the Peace Implementation Council (PIC[5]) expanded the executive and legislative powers of the Office of the High Representative (OHR), the internationally led organization responsible for implementing the GFAP. At the 1997 PIC conference in Bonn the High Representative was granted so-called 'Bonn Powers', to implement any law considered necessary for the implementation of the GFAP while removing any official (elected or otherwise) that was deemed to be obstructing this process.

Bonn Powers have presented a challenge to processes of democratization in BiH, and have led David Chandler (1999) to suggest that intervening agencies are 'faking democracy' where international supervision denies the Bosnian population the opportunity to participate in the political process. But this assertion of fakery suggests there is an 'authentic' form of democracy against which the political process in BiH may be assessed. While forms of political intervention have often sought to limit domestic political decision making, particularly when it contravenes the GFAP, it is not so straightforward as to constitute a counterfeit of Western models. Instead, as Toal and Dahlman (2011) suggest, we need to focus on the forms of institution and agency (the 'entrepreneurial field') that have been cultivated through processes of international intervention.

The establishment of the state is, then, at once a process of securing territory under the rule of a single administrative authority and the concurrent process of elevating that authority as somehow 'above' the society within which it operates. The GFAP created the impression of secure borders for BiH, through the signatures of the leaders of neighbouring states (most notably Croatia's Franjo Tuđman and Serbia and Montenegro's Slobodan Milošević). This imagined territorial security enabled the performance of many of the international signifiers of statehood, including the establishment of formal diplomatic relations, the creation of unified border controls and the acquisition of a seat at the UN General Assembly. But state existence came at the price of a compromise regarding the relationship between imagined ethnic identity and political space. As David Campbell (1998a, 1999) has illustrated, the GFAP retained the concept of a unified BiH while partitioning the state into ethnically attributed areas, in particular the two Entities of the Republika Srpska (RS) and the Muslim-Croat Federation. This act enshrined what Asim Mujkić (2007) refers to as the ethnopolitics of Dayton, where ethnic identity became endorsed as the primary affiliation of political participation.

The compromises at Dayton have had profound effects on the ability to convey the second sense of the state: that it operates 'above' a given society. State 'verticality' is a product of uneven power relations, where state agencies have the capability to act coercively and shape the behaviour of individuals and institutions (see Ferguson and Gupta, 2002). Such coercive force takes many forms, as exemplified by Bougarel et al. (2007), who explore a range of settings through which ideas of Bosnian statehood have been conveyed and resisted, including acts of commemoration, forms of justice and experiences of everyday urban life. While the assertion of Bonn Powers by the OHR suggests a form of executive authority for international representatives, the anthropological work of Bougarel et al. (2007) illustrates the forms of social and cultural practice that shape the possibility of implementing new forms of Bosnian statehood.

In the following chapters I build on this perspective to examine state building at both a local and national scale in BiH, exploring in particular how the distinction between state and society has been practised following the signing of the GFAP. Attempts to foster democracy in BiH have sought to bolster local civil society through funding autonomous NGOs, but despite pockets of success (Fagan, 2005; Ivanić, 2005) the effects have been to foster dependency on Western donors (Bieber, 2000, 2005) and promote a narrow interpretation of what constitutes democratic participation (Belloni, 2001). This focus draws attention to the ambiguous role of civil society in BiH where such organizations have been celebrated by both international and domestic political elites as evidence of the democratization of BiH society, while their agendas have often been shaped by the availability of funds or the nature of new regulatory obligations. Rather than acting as a focal point for political participation, I argue that cultivation of civil society expresses a wider desire to distinguish between state and society. Where in practice the lines between state and society are blurred to the point of indistinction, the enshrinement of civil society suggests an elevated state operating 'above' an arena of democratic participation. It is through a study of state as an improvisation, comprising both performance and resourcefulness, that the political effects of such ideas of the state come to the fore.

The approach taken over the following six chapters provides an insight into the dynamic nature of state practices in BiH as changing political contexts have established new priorities for both international and domestic politicians. The failed attempt to renegotiate the constitution of the BiH state in October 2009 illustrates how fundamental questions regarding the nature of BiH continue to be open to public debate (see ICG, 2009). Just as the GFAP was able to accommodate a range of divergent political interests, its enactment as a political framework has set the context for a variety of ideas of the state to coexist. For example, one of the central geographical imaginaries of changing statehood in BiH relates to the enrolment of ideas of Europe into political debate, where actors from across the political spectrum seek to bolster their positions through evidence of their European credentials (Ó Tuathail, 2005). But despite the consensus among international and domestic politicians concerning the virtue of the European project, studying the practice of closer European collaboration illustrates a range of underlying state projects.

1.3   Critical Geopolitics and Qualitative Research Strategy

Studying the state as an improvisation has a series of methodological implications. As suggested above, this approach to the state comes from a critical geopolitical perspective, a style of analysis that emerged from a desire to question the neutrality of geographical claims in the actions of political elites. In particular, the work of Gearóid Ó Tuathail (1986, 1996) has been at the forefront of analysis of conceptions of danger within Cold War US policy debates, where designations of the risk posed by the Soviet Union and its allies led to increased militarization of US foreign policy. Rather than seeing spatial designations of threat as innocent attempts to uncover certain geographical truths, Ó Tuathail's perspective foregrounds the productive nature of such actions: they simultaneously describe and produce a particular understanding of the world. In order to theorize this process Ó Tuathail and Agnew (1992) draw on the work of Michel Foucault (1980) to examine geopolitics as discourse, a means of exploring how political elites validate their understandings of the world through representations of global space. This approach centres attention on the forms of idiom and imagery through which designations of hierarchy and difference are geographically expressed.

Merje Kuus's (2007) exploration of geopolitical discourses of NATO expansion in Estonia provides an example of the implications of a critical perspective for questions of research strategy. Kuus's study employs a methodology of discourse analysis, where the researcher is not attempting to uncover hidden truths but rather to explore 'the persistent assumptions, themes, and tropes that both enable and constrain political debate and political practice' (Kuus, 2007: 9). Discourse, within this framework, is not restricted to speech, but r ather reflects the wider context that allows certain speech acts to appear feasible and logical and others unfeasible and illogical (ibid.; see also Müller, 2008). Toal and Dahlman (2011: 12) endorse this approach, defining geopolitics as a 'culturally embedded practice operating across networks of power' which needs to be approached as 'a field of competing political constructions vying to describe the conditions within which states operate and what normative strategy best realizes state and national interests'.

A critical approach to geopolitics has profound implications for understandings of research objectivity. Toal and Dahlman's emphasis on competition identifies the need to explore geopolitics as a dynamic set of practices, as individuals and institutions attempt to perform particular understandings of space. Moving from the abstract to the specific, this raises the question of where and when such geopolitical practices take place and how they may be apprehended. Traditional approaches to critical geopolitical analysis have privileged policy texts, official documents or newspaper sources as sites of analytical interest (see Ó Tuathail *et al.*, 2006). As a response to this work, feminist scholars have sought to provide more situated accounts of geopolitical practice that focus on the body, the home, the street and the workplace as varied sites of gendered geopolitical knowledge production (Dowler and Sharp, 2001; Hyndman, 2003). This perspective forces us to abandon a desire for a 'god trick' (Haraway, 1988), where the researcher proclaims an elevated position from which to articulate the nature of geopolitical realities. Instead, it highlights the positionality of the researcher as an active social agent that shapes the outcome of the research process (see Rose, 1997).

Across the next six chapters of this book I will be presenting a critical geopolitical perspective of statecraft in BiH, attentive to concerns regarding qualitative method and positionality. The research draws on four periods of residential fieldwork in BiH (2002–03, 2005, 2007 and 2009) and is informed more broadly by working for an NGO in BiH between 1999 and 2000. The switch from NGO worker to NGO researcher was initially difficult to navigate. My privileged access to audiences with international organizations led to a number of NGOs requesting advocacy on their behalf. These experiences reminded me of Nancy Scheper-Hughes's (1992) study of violence among street communities of North East Brazil. Prior to her research she had worked in these communities establishing a crèche and assisting with the immunization of young children. When she returned to Brazil to conduct fieldwork, she described how local people became resentful of her lack of action and help (Scheper-Hughes, 1992: 16). This only emphasized the importance of being explicit about the purpose of my research, that I was not affiliated to a donor organization or the UK government. As I discuss below, the changing research position underscored the significance of dissemination events designed to engage research participants in the process of drawing conclusions and building theory.

The research has used a qualitative, interpretive strategy, allowing the development of an understanding of the practices of the state in BiH, how they are shaped by human agency and how they have evolved through time. The aim of such methodologies is to explore the practices of everyday life to reveal 'the processes and meanings which undergird social action, and which enable order to be reproduced and sometimes challenged' (Herbert, 2000: 551). In the early research visits I was based in the northern Bosnian town of Brčko and in more recent visits in Sarajevo, Mostar and Bijeljina. The majority of the research participants have been members of local NGOs, community associations (*mjesne zajednice*), local state agencies and international organizations. The principal method of data collection was extended interviews and participant observation of NGO and state practices. Interviews, whether in English or Bosnian, were recorded and transcribed and these transcripts then discussed with research participants during repeat interviews. Towards the end of the research period in Brčko I held a dissemination event, where members of NGOs, political parties, the local state and international organizations were invited to discuss some preliminary findings from the research. Across all periods of residential fieldwork I have kept a field journal, excerpts of which appear in the following chapters to illustrate contextual points.

This methodology seeks to build on Megoran's (2006: 622) call for greater emphasis within political geographic inquiry on 'people's experience and everyday understandings of the phenomena under question', where political outcomes are understood as produced in the friction and interplay of social encounters. *The Improvised State* will therefore provide a grounded account of the nature and outcomes of Bosnian state practices since the GFAP. The utility of this approach is not simply specificity or empirical detail, but rather to provide a framework for understanding attempts to build state capacity in other settings at other times. Improvisation offers a means through which the production, reception and resistance to nascent state practices may be observed and theorized. Such an approach draws into sharp focus the limits of international intervention, the anti-democratic mechanisms that such processes can put in place and the significant scope for resistance to emergent state effects.

1.4   Structure of the Book

Chapter Two develops the theoretical framework of the book. The chapter opens with a review of literature that has sketched an anti-foundational account of the state, drawing particular inspiration from the work of Philip Abrams (2006 [1988]), Michel-Rolph Trouillot (2001) and Tim Mitchell (2006). As discussed above, this work has deviated from an understanding of the state as a pre-existing political and territorial formulation that may be studied in isolation from wider social, cultural and political processes. The challenge posed by the state, or 'the state effect' in Mitchell's (2006) terms, is to identify how the idea of a detached and authoritative entity is established and reproduced. *The Improvised State* explores this process through an understanding of the state as a result of a set of improvisations. As discussed above, improvisation captures a dual process: both *performance* and *resourcefulness*. When drawing on musical examples such as free-form jazz, scholars have stressed the lack of planning or prior stipulation and cited improvisation as acting 'on the spur of the moment' (see Weick, 1998). In this sense improvisation orientates our attention to the performed nature of social and cultural life as it is enacted through the intent of the performer. In contrast, accounts of social improvisation have tended to explore improvisation as the ability to 'make do' with available materials. This approach is developed through the work of French structuralist anthropologist Claude Lévi-Strauss (1972) and his notion of *bricolage*, defined as the use of pre-existing materials and repertoires in new configurations of social and cultural practice. But the utility of Lévi-Strauss's vocabulary is constrained by his silence on questions of power or dominance (see Werbner, 1986). As a corrective, Pierre Bourdieu's concepts of *habitus* and *capital* (1984) have sought to introduce a sense of agency and constraint into understandings of bricolage. In doing so, Bourdieu's approach alerts us to the role of structured improvisations in reproducing particular elite class advantage through social and cultural practice. This dual framework of analysis – both performance and resourcefulness – illuminates contemporary state practices in BiH and allows us to understand how particular political perspectives have solidified in the post-conflict period.

Chapter Three develops the theoretical foundations outlined in Chapter Two, through an exploration of the geopolitical histories of BiH. As set out above, it is not the purpose of this book to establish a definitive history of the causes or outcomes of the conflict. Instead, I want to investigate how geographical arguments, ideas and concepts have become central to under-standings of the violence. In itself this is not new; geographers and scholars from affiliated disciplines have been examining the geopolitics of the con-flict for some time (see Campbell, 1998a; Ó Tuathail, 1996, 2002; Simms, 2001; Toal and Dahlman, 2011). But within this chapter I build on this work to explore how plural understandings of space and power have con-tributed to present-day improvisations of the state. I argue that there are three prominent spatial narratives to the history of violence in BiH: the fault line, the barrier and the vortex. I explore each of these geopolitical frames in turn, and examine some of the ways in which they have been adopted within cultural and political practices. The key point here is that these spa-tial stories contradict each other, since they envisage different ideas of the relationship between political communities and space. Perhaps the clearest example of this tension is the GFAP, which attempted to incorporate aspects of all three spatial histories.

Such plural and conflicting performances of the state will be explored through an examination in Chapter Four of state-building strategies in the period after the signing of the GFAP. Using the localized example of Brčko District this chapter explores how improvisations of the state are ephemeral moments that attempt to claim permanence. The idea of the state may be dynamic and plural, but observations of its performance suggest stability, durability and stasis. This argument is made through an examination of three improvisations of the state: stability, security and neutrality. Each of these sections examines how political elites sought to convey a sense of a coherent local state, drawing in particular on the symbolic resource of the ability to create law. While the empirical focus may be Brčko District, the story that unfolds illustrates a common refrain in state-building contexts: plural performances of state authority coexist across the same territory. The enactment of Brčko District as a local state institution required the unifica-tion of laws to change individual understanding of the nature of statehood, to make people 'think' in terms of the new idea of the state.

Developing the discussion, Chapter Five explores the nature of Bosnian state improvisation in the example of democratization. Democratization has been one of the central discourses of international intervention in BiH (Bose, 2002; Campbell, 1998a). This chapter explores how this discourse has been performed in terms of both the spectacular nature of presidential and local elections (including the performance of voting and the symbolic nature of election posters) and the more substantive conceptions of democratization relating to attempts to foster civil society organizations. This discussion develops the application of Bourdieu's ideas to explore the ways in which international donors and regulators have conferred legitimacy on particular styles of democratic participation, while sidelining or legislating against other practices. I argue that systems of funding and regulation have gentrified civil society, producing a set of compliant organizations that entrench particular understandings of state power within BiH society. It is through this discussion that the resourcefulness at the heart of state improvisation is brought to the fore, as regulating bodies employ historical precedents (such as the Yugoslavian concept of the *mjesne zajednice*, or local community associations) in order to foster associative life. In turn, newly formed NGOs have exploited a range of repertoires in order to gain access to funding and respect, from improving English language skills to developing links with hard-line nationalist political parties such as the *Srpska demokratska stranka* (Serb Democratic Party or SDS). This discussion highlights the contradictory nature of state improvisation: while it provides space to challenge the often anti-democratic nature of international intervention, it illustrates that the resulting critiques are not necessarily emancipatory. They are instead offering differing visions of state performance, enshrined in particular understandings of identity and territory based on ethno-national affiliation.

Chapter Six explores how attempts to establish instruments of transitional justice have shaped the improvisation of the Bosnian state. The chapter focuses in particular on the process through which transitional justice has been 'localized' in BiH since 2005 through the establishment of a War Crimes Chamber (WCC) of the Court of Bosnia and Herzegovina. This institution is due to take over the competencies of the International Criminal Tribunal of the former Yugoslavia (ICTY) at The Hague when its mandate ceases at the end of the trial of Radovan Karadžić. The chapter examines how this reorganization of judicial functions has reshaped the practice of state improvisation in BiH. In particular, it focuses on the mechanisms through which the WCC has attempted to build its legitimacy with Bosnian citizenry through its outreach programmes. This illustration provides evidence of the competing notions of the Bosnian state that have been performed since 1995, from the GFAP's pragmatism to the idealism offered by claims to retributive and restorative justice. This tension is explored through the everyday responses of civil society organization to initiatives to bring war criminals to trial.

The final substantive chapter explores how understandings of state improvisation assist analysis of BiH's struggle for accession to the EU. This discussion illuminates the variety of different interpretations of Europe that have circulated within political discourses in post-GFAP BiH. Rather than standing as a signifier of a post-territorial or cosmopolitan political future, it is argued that Europe is mobilized as a means through which to bolster a range of state projects within BiH.

The conclusion provides a summary of the key arguments made across the preceding seven chapters. In scholarly terms the focus on improvisation draws attention to the significance of the performance of sovereignty, a fleeting, dynamic and contested set of practices that are always incomplete, evolving, and constrained by available resources. In so doing, the analysis highlights a set of policy implications, relating to the effectiveness of state performances and their ability to foster democratic and shared understandings of political practice. The objective of these discussions is to move beyond the illumination of contradictions of Bosnian statehood towards an understanding of the political effects of such contradictions as they are played out in BiH. By highlighting the lived experience of BiH statehood this discussion will contribute to analysis within human geography and affiliated disciplines that has explored the nature of the contemporary state and the moral underpinnings of international intervention in the sovereignty of post-conflict states.

Notes

1   A note on abbreviations and terminology: the state of Bosnia and Herzegovina will be abbreviated to BiH for the remainder of the book. I try to refer to ethnic groups (Croat, Serb, Bosniak) sparingly (for the reasons given in the opening pages of Chapter Three), but when I do, I acknowledge that these are imperfect labels that are contested by both those they encompass and external commentators. They are, however, labels that are common in everyday life and political discourse in BiH.

2   These figures are derived from the 1991 Yugoslav Census, which may not be an entirely reliable picture of Yugoslavia's demography, not least since a number of groups advocated boycotting the poll or were placed under duress to withdraw participation (see Bennett, 1995).

3   Again these figures are drawn from the 1991 Yugoslav Census and should be approached with caution for the reasons stated in n. 2.

4   The presentation of the conflict as a 'civil war' is further complicated by the presence of Croatian military forces on BiH territory, particularly towards the end of the war (see Silber and Little, 1996).

5   Founded at a conference in London in December 1995, the PIC is a coordinating body representing 42 states and (currently) 13 international organizations. The PIC oversees the work of the OHR and its role in implementing the civilian aspects of the GFAP.

Chapter Two

The Improvised State

The framed letter hanging above his desk stood out starkly against the bare walls. I could make out the logo of the Office of the High Representative (OHR), but I couldn't read the contents of the letter (nor would it be a good idea since we were sitting in an interview). So I asked the Mayor, 'What is the purpose of the letter?', The mood of the interview changed dramatically; where previously the discussion of civil society in Brčko had been soliciting a muted response, the Mayor seemed animated and engaged. 'The letter?', he said, rising to his feet. 'It is the letter sent to me by the OHR to announce my appointment as the Mayor of Brčko District.' At this point he took the letter from the wall and began to 'act out' the following scene, describing his appointment: 'It was quite an occasion', he said. 'Three of us were in a room and the District Supervisor [the leading international representative in Brčko, see Chapter Four] entered and pointed to each in turn: you will be Mayor, you will be Speaker of the Assembly and you will be Deputy Speaker. So he had recruited a Serb, a Bosnijak and a Croat.'

Notes from field journal and interview with the Brčko Mayor,
8 May 2003

Theatricality is at the heart of the state. This exchange, taken from an interview with a local government official in Brčko District, illustrates a number of ways in which this theatricality can be conveyed, both materially (in the case of the letter) and performatively (in the case of the actions of key individuals). It also illuminates the range of institutions that are implicated in performing the state. The OHR, the leading international organization in BiH tasked with implementing the General Framework Agreement for Peace (GFAP), sent the letter as a symbol of the authority placed in the Mayor, thus underscoring the significance of this international institution in the accomplishment of BiH statehood. The Mayor then hung the letter on the wall, in order to demonstrate the trust that had been placed in him and his legitimacy in holding office. He then literally performed the scene where he had been appointed, and placed it in the context of the ethnic politics of Dayton BiH. But each performance also produced an audience: from the three politicians observing the District Supervisor's appointment routine, through to my own role as a researcher observing the letter and subsequent explanation. While the state may conjure a sense of a static political institution, its existence is predicated on repeated ephemeral performances of its sovereignty.

This chapter sets out the theoretical perspective of the book. It opens with an account of the struggle faced by scholars over the past three decades seeking to theorize the state. This discussion captures the profusion of studies that have sought to question the supposed solidity and unity of the state. Inspired variously by Marxist, feminist and post-structural approaches, this work has introduced a new vocabulary of the state using a language of magic, fantasy and affects to identify the plural means through which the state secures its legitimacy as a form of rule. This argument has two inter-linked implications. First, it critiques strands of political science that have assumed that the state stands as the backdrop to political life. This assumption is inverted within recent critical scholarship to turn the gaze onto the mechanisms, materials and infrastructures that secure this assumption as the starting point of political analysis. This leads to a second implication of this scholarship: that scholars have sought to challenge the ontology of the state, seeking to theorize the state as an idea rather than as a political or material reality. This approach has led to a range of studies examining how particular ideas of the state are conveyed through everyday life and made meaningful through specific practices and traditions. One of the key frame-works for understanding the public expression of state ideas has emerged from theatrical metaphors and, in particular, from ideas of performance.

Following this engagement with state theory, the chapter moves on to explore the varying ways in which conceptions of performance have been incorporated into understandings of political practices. The distinction between performance and performativity is examined, focusing in particular on the implications of this schism for understandings of human agency. One of the main challenges of adopting theatrical metaphors is the underlying sense of a pre-given scripting of individual action that defies a sense of spontaneity and invention. In order to recover these traits within an understanding of the state as socially performed, the book incorporates understandings of performance with work exploring resourcefulness and adaptation to argue for an understanding of the state as improvised. This final section defines improvisation as a practice of 'performed resourcefulness' that simultaneously foregrounds individual and collective capabilities alongside the structuring effects of existing and emergent material and symbolic contexts. Significantly, this definition moves away from understanding improvisation as simply 'free will' and moves towards a more anthropological definition based on capacities, practices and implications. The final section of the chapter adopts this understanding of 'performed resourcefulness' for conceptualizations of the state.

2.1   The State Idea

Critical accounts of the state have struggled with a paradox. Attempts to explain the nature of statehood have seemingly required an assumption that the state exists as a knowable social and political entity. As Peter Bratsis (2006: 9) has explained, state theorists have been hampered by assuming the existence of the state, proceeding 'as if' the state 'was indeed a universal *a priori* predicate to our social existence rather than the product of our social existence'. Part of this challenge is a reflection of the success of the state to present itself as a universal political and social reality, a form of disinterested domination that is set apart from other political and social practices (Trouillot, 2001). While the terminology involved suggests a singularity ('the state'), what is signified is a more complex relationship between political power and social life. The assumption of state existence is also a reflection of the very cumbersome nature of actual existing state bureaucracies, where individuals, agencies and effects are diffused beyond state territorializations (Abrams, 2006 [1988]: 114). Consequently, the empirical challenge of apprehending the state has cultivated a form of theoretical abstraction that reproduces a sense of the state as a concrete institutionalization of political power over space.

One of the implications of theorizing the state from this abstract perspective is that it reproduces a sense of the state as separate from, and acting upon, wider society. Painter (2003) suggests that the tendency to separate state and society stems from traditional disciplinary perspectives that have often sought to explore either the state (within political science, international relations or political geography) or society (within sociology, anthropology, development studies or social geography). This sense of separation is further enshrined in the work of Max Weber, the most regularly cited scholar on the state. Weber (1958: 78) defines the state as a 'human community that (successfully) claims the monopoly of legitimate use of physical force within a given territory'. There are two main consequences of using this definition. First, and reinforcing the point above, it gives rise to a sense of the state as guaranteeing the security of a wider population (society) through military force. Second, it places legitimacy at the heart of the achievement of statehood: that the monopoly of physical violence has to have gained some recognition as a legitimate political act. It is less clear in whose eyes this legitimacy may be sought, in particular whether this refers to the recognition of an imagined society, other states or intergovernmental institutions (or all three).

Critical approaches to understanding the state have often reproduced the concept of state and society as separate spheres. This tension is the product of a desire to deconstruct and decentre the practices of the state as diffused through society while simultaneously retaining a unified object of critique. For example, Marxist scholarship has sought to explain how the state sustains and guarantees fundamental relationships of capitalism (Harvey, 1976; Painter, 2003). But Marxism has never been simply a project of theorization and reflection; it has at its core a political project of intervention and transformation. Philip Abrams (2006 [1988]: 118) sees this as producing particular challenges for Marxist perspectives, which have presented the state as both an abstract entity that enters into the reproduction of relations of production (and therefore class struggle) and simultaneously a concrete set of institutions that are the object of transformative politics. This tension is illustrated in the work of Nicos Poulantzas (1978), who seeks to understand the ideological role of the state, while refuting the (then) emerging post-structural critiques of the state power advocated by scholars such as Michel Foucault and Gilles Deleuze (see also Jessop, 2008). Poulantzas (1978: 28) challenges understandings of state power as diffused through micro-situations while simultaneously noting the significant role played by social imaginaries in reproducing state power:

> The State cannot enshrine and reproduce political domination exclusively through repression, force or 'naked' violence, but directly calls upon ideology to legitimize violence and contribute to a consensus of those classes and fractions which are dominated from the point of view of political power. Ideology is always class ideology, never socially neutral.

The paradox remains, then, between a critical stance that challenges the ontology of the state as a coherent set of institutions and continued reliance on the distinction between state (as political actor) and society (as subjects of state violence).

Rather than seeking to theorize the state through its actions, recent scholarship has sought to understand the state through its effects. In this sense orientation has shifted from aspects of Weber's definition identifying a tangible physical entity (a human community occupying a distinct territory) towards the more ideological question of the conferment of legitimacy (over the exercise of physical violence). As Bratsis has pointed out, 'if force must be legitimate in order for the state to exist, then the cognitive and affectual processes that create this legitimacy must be of primary interest' (Bratsis, 2006: 14). This insight shifts attention away from thinking about the specific material attributes of the state and towards the subjective processes through which the state is secured as a stable truth. Abrams's (2006 [1988]) account is an early attempt to advance this approach, as he advocates a move away from isolating the institutional reality of the state to think of the state instead as an 'idea'. For Abrams (2006 [1988]: 122) the function of the state idea is not neutral or transparent, rather the idea 'mis-represent[s] political and economic domination in ways that legitimate subjugation'. He continues (2006 [1988]: 123):

> The state is [...] in every sense of the term a triumph of concealment. It conceals the real history and relations of subjugation behind an historical mask of legitimating illusion; it contrives to deny the existence of connections and conflicts which if recognized would be incompatible with the claimed autonomy and integration of the state.

Abrams's work has had both theoretical and empirical implications for the subsequent study of the state. In theoretical terms scholars have sought to challenge the production of state knowledge and question the forms of subjectivity that it entails. This focus on knowledge and subjectivity has been enriched through recourse to Foucault's detailed historical expositions of the relationship between power and knowledge in the operation of rule. In particular, Foucault's concept of governmentality has been central to critical interpretations of state discourse (Jessop, 2006, 2008; Mitchell, 1991, 1999; Painter, 2006; Weber, 1998). Governmentality focuses attention on the diverse political rationalities of government, on its 'technologies', and on the considerable intellectual labour involved in bringing into being the things, people and processes to be governed (Painter, 2002: 116). In relation to the state, Foucault described governmentality as the 'tactics of government which make possible the continual definition and redefinition of what is within the competence of the state' (Foucault, 1991: 102). By examining government in these terms the state no longer stands as an unquestioned *source* of power but, rather, as its *effect* (Marston, 2004: 4). Governmentality has become shorthand for explaining how state power is vitalized, as tactics and technologies are deployed to control, subdue and oppress the citizenry (Butler, 2004). Scholarly attention has consequently shifted away from a quest to trace the functions of the state, and towards an interest in the effects of the state idea. For Timothy Mitchell (1991: 81) such effects are diffuse and difficult to grasp, but it is precisely this imprecision that has been the 'source of its political strength as a mythic or ideological construct'. So instead of dismissing the state and seeking a coherent and tangible alternative category (such as government or political party), Mitchell suggests that it becomes more pressing to understand state effects.

This theoretical move has had significant consequences for how the state is understood empirically. For example, Mitchell (1991: 82) suggests that understanding the state as an idea collapses the distinction between conceptual and empirical realms, since it is within social processes and interactions that the state establishes its existence. Painter (2006) makes a similar point, arguing that to understand the state requires engagement with the prosaic and everyday processes through which the state is realized. Indeed, Painter uses the term 'statization' to draw attention to the dynamic and socially embedded process through which the state comes into being. The wider consequence of this focus on empirics has been the adoption of these conceptualizations of the state by political anthropologists. Scholars such as Trouillot (2001), Ferguson and Gupta (2002), Sharma and Gupta (2006), Navaro-Yashin (2002), Taussig (1997) and Corbridge *et al.* (2005) have sought to employ ethnographic techniques in order to understand the social and cultural processes through which state ideas are reproduced. For example, Ferguson and Gupta (2002) explore the reification of the state across two spatial manoeuvres: verticality and encompassment. State verticality refers to the 'central and persuasive idea of the state as an institution somehow "above" civil society, community and the family' (Ferguson and Gupta, 2002: 982). This topography of power is reproduced through the rhetoric of 'top-down' policies, layers of bureaucracy or even the notion of the 'head' of state. Ferguson and Gupta (ibid.) go on to use the term encompassment to refer to the significance of the state idea in an 'ever widening series of circles that begins with the family and local community and ends with a system of nation-states'.

Focusing theoretically and empirically on the state as an idea is productive in two ways. First, conceptualizing the state as an idea draws attention to the continual processes through which the state is reproduced. This is not exclusive to cases of state building, though it may be that attempts to convey ideas of the state are more explicit in moments where alternative state ideas coexist. This approach serves to illustrate the everyday mechanisms that reinforce particular understandings of the state, and the ways in which ideas about the state acquire legitimacy within the consciousness of citizens. As we have challenged the ontological foundations of the state as a distinct material entity, so we must turn our attention to the performances that reify the state on a daily basis. Second, rejecting clearly delineated boundaries to the state draws in a wide array of individuals and groups in the constitution of the state idea. Mitchell (1991: 93) makes clear that as we abandon the state as a free-standing agent issuing orders, we also need to question 'the traditional figure of resistance as a subject who stands *outside* the state and refuses its demands'. As we will see through the book's later chapters, the geometry of power and resistance within state building is complex and plural. Often alternatives to the state exist both within and beyond established discourses of the state, and it is this coexistence that illuminates potentially transformative political pathways. We return to these debates concerning human agency towards the end of the chapter, but first we must explore the nature of state theatricality in more detail through a discussion of the diverse ways in which performance has been utilized as an analytical tool.

2.2   Performance and Performativity

The theatre has provided the social sciences with a rich set of terms and metaphors through which to narrate the complexity of social and political life. Contemporary writing across disciplines as diverse as business studies, anthropology, geography, political science and sociology relies on a vocabulary of actors, audiences, mimicry and performance. Indeed, so prevalent is theatrical language that scholars often assume the content of these terms to be self-evident. But the common adoption of dramaturgical vocabulary masks fundamental disagreements about what theatricality means and the possibilities it affords towards understanding phenomena under observation. In part, such disagreements reflect common schisms in disciplinary perspectives, but it is also a consequence of the development and extension of theatrical metaphors over the past seventy years.

One of the early utilizations of dramaturgical metaphors emerges in one of Erving Goffman's attempts to understand human interactions as performances in *The Presentation of Self in Everyday Life* (1959). This work explored everyday life through theatrical relationships such as performer and audience and front stage and back stage (or, in Goffman's terms, the 'front region' where a performance is enacted and 'back region' where a performance is prepared). His work centred on an investigation of individual intentionality, exploring 'the way in which the individual in ordinary work situations presents [their] activity to others' (Goffman, 1959: ix). While the emphasis on intentions conjures a notion of the unbridled agency of the individual, Goffman's work stresses constraints placed by the 'needs of the presentational self' (Rawls, 1987: 136). This explanation highlights individual attempts to reproduce established social orders, though Goffman drew this into a collective framework through the notion of 'teams' that cooperate to present a particular performance to a given audience. In this work theatricality serves as an analytical tool for describing social interaction, a means through which the nature of social affiliations, rituals and groupings may be described and illuminated. Goffman's analysis provides a framework for exploring the intimate social dynamics of workplaces and homes (Crang, 1994; Enck and Preston, 1988), while also providing analytical tools for confronting larger-scale phenomena such as international intervention (Andreas, 2008) and democracy building (de Souza Briggs, 1998).

Goffman's account of performance helps to illuminate the plurality of social life, where intentions are formulated and then enacted both individually and collectively in order to preserve a particular social ordering. For example, Andreas (2008) uses Goffman's concepts of 'front' and 'back' regions in order to illustrate the public and private manoeuvrings of local and international agents in Sarajevo during the Bosnian war. Andreas (2008: 8) writes: '[w]hile their front-stage behaviour was often carefully staged and choreographed for various audiences [...] backstage there was greater room for improvisation and deviation'. The relationship between front and back stage does not equate to a straightforward division between performance and preparation, as Andreas (ibid.) notes:

> [W]hile the Bosnian Serb leadership played the front stage role of orchestrating the siege of the city under the official banner of ethnic grievance and animosity, backstage they profited from the siege through clandestine business dealings that included cross-ethnic economic exchange across the frontlines.

Goffman's framework is thus utilized to allow the reader to explore the dissonance between the public iterations of political leaders concerning the siege and the simultaneous private collusion of the same officials with besieging military groups.

While Goffman's framework evokes the duplicity of social life, it has been criticized both for its content and for its philosophical implications. In terms of the content of Goffman's ideas, scholars have unsettled the spatial determinism of discerning spaces as 'front' or 'back' regions. For example, Crang (1994) sees the conceptual division of 'front' and 'back' regions as only a starting point towards theorizing individual behaviour. In his account of waiting tables in a US-themed restaurant Crang suggests that spaces had traits of 'frontness' and 'backness' rather than performing a singular function. This more textured account allows for temporal change, where regions may assume or simultaneously exhibit degrees of 'frontness' and 'backness'. In addition to this criticism of spatial determinism, Giddens (1984: 70) suggests that Goffman's analysis assumes motivated agents 'rather than investigating the sources of human motivation'. For Giddens, such an investigation would require reflection on the nature of the subconscious and the wider macrosociological forces that shape human encounters, two issues that are unexamined by Goffman. Thus while Goffman's account sees intentions shaped by the need to conform to social norms and mores (his notion of 'preserving the presentational self'), this says rather less about the formulation and reproduction of this symbolic ordering.

In addition to this analysis of the content of Goffman's framework there have been perhaps more profound criticisms of the implications of his theatrical approach for understanding the nature of subjectivity. As Gregson and Rose (2000: 433) noted, 'behind Goffman's analyses of interaction lies an active, prior conscious and performing self'. This approach separates the performer from the performance, rather than a more complex understanding of the interrelations between, or transcendence of, these two groups. This idea of separation points to a potential explanation for the endurance of Goffman's ideas within the social sciences. The continued reliance on Goffman's approach is perhaps less an indication of the relationship between theatre and social life and more a reflection of the theatrical nature of masculinist academic practice. There is symmetry between the social scientist's desire for a bounded site of inquiry and Goffman's conceptualization of distinct social performances. In addition, the notion of performance allows the scholar to assume the roles of director and audience, with a privileged knowledge of the scripts, costumes and comportment of the 'actors' involved. This directorial role reflects the supposed licence of the academic to set the terms of social analysis. For example, it is the academic who first discerns the distinctions between front and back regions and then goes on to gaze across these sites to adjudicate social practices. The notion of social life occurring on a 'stage' also allows a conceptual distance to be drawn between the empirical realm and the scholarly practice of analysis and theory building (reflected in the language of being 'in the field'). Where the ubiquity of theatrical metaphors points to a desire to understand the production of social life, they also (often inadvertently) reproduce this masculinist fantasy of scholarly practice. In so doing, the power relations that structure both the performance and academic analysis are often occluded within technical language of distance and expertise.

Over recent years cultural geographers have sought to recover notions of performance from Goffman's theatrical approach. Rather than traditional preoccupations with texts and representation, scholars have sought to enrol a language of performance in order to communicate a sense of cultural practice as embodied and unfolding (Nash, 2000; Thrift, 2003). In moving beyond a theatrical notion of performance, feminist and queer theorists have drawn on the post-structural accounts of Jacques Derrida to develop a notion of 'performativity'. This broad range of perspectives has sought to question the concept of a singular, stable subject and explore instead the forms of power relations and practice that bring certain subjectivities into being (see Müller, 2008). Judith Butler's (1990) study of gender identity is one of the foremost examples of this style of post-structural analysis. Butler's approach seeks to advocate a linguistic, as opposed to theatrical, approach to performance, analysing how particular categories and identities become embedded in the fabric of social life. Using the term 'performativity', Butler argues that linguistic iterations bring certain categorizations (such as 'male' and 'female') into being, reflecting the assertion that 'discourses constitute the objects of which they speak' (Bialasiewicz et al., 2007: 406). This is a vital point: within Butler's schema iterations are never just words; they summon into existence and naturalize particular concepts, categories and modes of being.

Butler's discursive approach illustrates her debt to Foucault. By drawing on Foucault's exploration of the relationship between power and knowledge,

Butler espouses an anti-foundational approach to sexual identities, suggesting that 'genders can neither be true nor false, but are only produced as the truth effect of a discourse of primary and stable identity' (Butler, 1990: 136). This move sought to challenge existing feminist political and intellectual perspectives that had been founded upon the category of 'women' as a starting point for transforming masculinist hegemony. Rather than adopting a form of strategic essentialism, Butler argues that such categorizations 'effect a political closure on the kinds of experiences articulable as part of feminist discourse' (Butler, 1997: 248). Butler turns to psychoanalytic accounts, in particular of Freud and Lacan, to explore the production of masculine subjectivity. One of the significant outcomes of this analysis is the absence of symmetry between masculine and feminine subjectivity. It is not that Butler presents a simple disequilibrium, but rather argues that the notion of the subject is a 'masculine prerogative within the terms of culture' (Butler, 1997: 249).

Butler uses the notion of performativity as a means of moving beyond discussions of subjectivity and thinking in terms of events and practices (see Thrift and Dewsbury, 2000). Consequently, the suggestion that gendered identities are performative shares with Goffman's notion of performance a focus on the enactment of particular styles of cultural practice. However, we can identify at least two important divergences in their understandings of performance. First, Butler's approach conveys a different sense of subjectivity, questioning the notion of a stable ontological subject that prefigures action (see Nayak and Kehily, 2006). As we have seen, Goffman's analysis of 'intentions' is suggestive of a conscious and detached subject possessing the agency to operate outside of prefigured constraints (see Gregson and Rose, 2000). Butler, in contrast, argues that there is no subject that 'precedes or enacts the repetition of norms' (Lloyd, 1999: 201). In contrast, 'the subject is the effect of the compulsory repetition' (ibid.). This account of subjectivity is far from straightforward, since it seems to deny both the transformative power of the individual and the absolute structuring force of existing social contexts. As such, Butler has been criticized for denying individual or collective agency (see Nelson, 1999). However, as Bialsiewicz *et al.* (2007) argue, Butler does not dispense with agency *per se*; rather, she seeks to use the notion of performativity to advance a more embodied and emergent account of the relationship between personal agency and social structures (see also Nash, 2000: 654).

The second divergence between Butler's linguistic approach and Goffman's theatricality relates to the potential boundaries that may be placed on performance. Within Goffman's schema, performances are presented as bounded acts that take place within a performance space (the 'front region') and therefore retain discernible temporal and spatial boundaries. In contrast, at the heart of Butler's account of performativity is a notion of repetition, where gendered identities are monitored and incited 'through a process of cultural reiteration' (Segal, 2008: 382). While her accounts of gendered identity call attention to individual performances of gender, these operate within a wider 'infrastructure of performativity' (Bialasiewicz *et al.*, 2007: 407). Since this infrastructure is an embodied set of practices and dispositions that reproduces dominant understandings of the world, it could not be bounded into an isolated performance; its sense is derived through its repetition.

The distinction between performance and performativity has been the subject of sustained academic debate, perhaps understandably given the blurred boundaries between the two (Gregson and Rose, 2000; Nash, 2000; Thrift, 2003). This reflects a broader conceptual uncertainty as to the possibilities for human agency within Butler's work (see Lloyd, 1999). Butler deviates from first-wave feminist concerns with gender equality and emancipation from patriarchal oppression. Rather than collective struggles towards normative ends, Butler advocates a form of political practice centring on parody. Since gender is a 'doing' (it is produced through repeated performances), it may be 'undone' through modifications of those performances. Butler argues that parodying existing gender identities, in particular through the use of drag, acts as a means through which the fabrication of gender may be illuminated (see Bell and Binnie, 2004). As Lloyd (1999: 209) explains, this involves a subtle process of 'resignification':

> the demand to resignify and repeat the very terms which constitute the 'we' (woman or lesbian) cannot be summarily refused but neither can they be followed in strict obedience. [...] The idea that occasional gaps open up within contemporary culture where norms (both dominant and annihilating) can be mimed, reworked, resignified and, of course reincorporated is suggestive.

It is through these 'occasional gaps', Butler argues, that dominant categorizations may be subverted. But this account leaves considerable ambiguity as to what constitutes a parodic performance and the real possibilities that these undermine oppressive or iniquitous patterns of power. Martha Nussbaum, a staunch critic of Butler's approach, argues that such an account of political activism equates to moral passivity, or simply 'waiting to see what we get' (Nussbaum, 1999: np). Following Nussbaum, it is apparent that Butler's approach does not give a clear indication of how or why particular subversive parodies are considered virtuous (a point also made by Lloyd, 1999: 210). There is an implicit sense in Butler's schema that any action that challenges dominant forms of categorization presents a social good. But parodies may just as easily undermine progressive politics through the destabilization of political collectivities that have been structured around specific identities.

Butler's ambiguity on politics represents both her strength and her weakness. Her writings provide a set of concepts that move beyond Goffman's rigid theatrical framework and illustrate the significance of the embodied nature of social life and the centrality of language in structuring thought. But the relationship between performance and performativity remains opaque and consequently provides less assistance in analysing empirical examples. Where scholars have attempted a more grounded engagement with Butler some of the political possibilities come to the fore (see Bialasiewicz *et al.*, 2007). This approach is illustrated in Gregson and Rose's (2000) examination of the practices of community art workers and car boot sales attendees. Within this work (Gregson and Rose, 2000: 441) performativity acts as a backdrop for performances:

> [...] performance – what individual subjects do, say, 'act out' – is subsumed within, and must always be connected to, performativity, to the citational practices which reproduce and subvert discourse, and which at the same time enable and discipline subjects and their performances.

But Gregson and Rose seek to extend and challenge Butler's account of agency by investing spatiality into her theoretical schema. Consequently, they do not see performance as simply reiterative of a singular subject position but rather as generative of multiple subject positions through the production of space, implying 'the possibility for slippage, subversion, disruption and critical reworking of power' (Gregson and Rose, 2000: 446). The authors use their empirical examples to present a 'messier' and 'more unpredictable' (ibid.) conceptualization of performance, which illuminates the collective and collaborative performances that exceed notions of the individual. This is a qualitatively different notion of subjectivity that operates beyond discursive reiteration and subversion to include the production of spaces through which power relations may be questioned. Thus this spatial understanding seeks to avoid an account of individual agency that implies the unquestioned reproduction of existing discourses, and views space as a means through which forms of political action may be enacted.

This empirical application of Butler's ideas presents a form of subjectivity that is more conscious of the machinations of power and consequently presents 'gaps' for reworking dominant discourses of community activism and neoliberal consumption. In this sense, this work is open to Lloyd's (1999) critique (levelled at Bell and Binnie, 1994) that it reinstates the voluntarist subject that Butler jettisons. But where the voluntarist subject is (at least partially) restated by Gregson and Rose, there is a residual uncertainty as to the motivations behind individual and collective performances, beyond a statement that they are 'bound up and enmeshed in very complex ways' (Gregson and Rose, 2000: 401) with the established knowledges which they cite. In order to draw on, and extend, this form of analysis, we need to return to the concept of performance and investigate further what it is to act. Butler's account helps us develop a language of performativity, illuminating how performances summon into existence identities, categories and social relations. But in order to understand the political possibilities inherent in this schema, we need a more nuanced account of how individual performances are constituted and enacted. This requires a clearer understanding of the symbolic economy within which performances take place. It is to this point that the chapter now turns.

2.3   Improvisation: Performed Resourcefulness

Improvisation has long been used as a term that defines some form of extemporaneous action within the theatre (Johnstone, 1981), music performances (Berliner, 1994) or business studies (Baker *et al.*, 2003). By conveying spontaneity in action, improvisation extends the notion of performance found within either Goffman's dramaturgical metaphor or Butler's more linguistic conceptualization. In both understandings of performance the ability to operate 'on the spur of the moment' is constrained, through either an assumption of precognition or the impossibility of individual subjectivity. Instead of mobilizing a language of performance, scholars have used the term 'improvisation' to describe moments of practice where the subject may utilize, subvert and reformulate existing discourses in order to produce a new and previously unseen event or practice. For example, in a study of entrepreneurialism Baker *et al.* (2003) use the concept of improvisation to provide a means of analysing the convergence of the design and execution of start-up companies. This approach orientates attention to the practice of performance as more than a reflection of existing discursive frameworks or circulating scripts, but as a moment of production and industry.

The concept of spontaneity lies at the heart of theatrical definitions of improvisation. This is illustrated by Frost and Yarrow (1989: 1), who define the practice of improvisation in the theatre as:

> the skill of using bodies, space, all human resources, to generate a coherent physical expression of an idea, a situation, a character (even, perhaps, a text); to do this spontaneously, in response to the immediate stimuli of one's environment, and to do it *à l'improviste*: as though taken by surprise, without preconceptions.

But this definition draws attention to more than spontaneity; it reflects the concept of improvisation as a multifaceted practice. In addition to the performance component of improvisation ('the physical expression of an idea'), Frost and Yarrow illustrate two further elements: spontaneity and response to the immediate context. These two interlinked aspects of improvisation illuminate a more complex subjectivity than simply 'acting on the spur of the moment'. Rather than a straightforward expression of free will, the process of improvisation requires the negotiation of pre-cognitive and cognitive impulses that reflect individual judgements of appropriate practice. These judgements are set within the context of an environment of limited possibilities, in terms of language, bodily expression and external resources. Therefore improvisers have to work with what is available and order this in a way that suits their immediate preferences. Improvisation is not simply performance; it is also a practice of resourcefulness.

Thinking of improvisation as a form of resourcefulness seems to work in opposition to spontaneity, invoking a notion of calculated practice that deviates from an expression of free will. But such a tension presupposes that there exists an authentic form of spontaneity that exists beyond material, ethical or political considerations. This tension has been explored by drama theorists such as Keith Johnstone (1981), who attempts to cultivate a form of 'learnt spontaneity' in his students. He argues that in order to improvise you must shed the layer of education and enculturation that prevents the expression of 'true' spontaneity. Johnstone (1981: 119) suggests that when you act spontaneously, you reveal 'your real self, as opposed to the self you've been trained to present'. Within this schema improvisation requires un-learning the traits assumed through formal education and retraining the actor's mind and body. Rather than striving for originality (which Johnstone perceives to be a false quest), cultivating this style of improvisation requires reflecting the 'true' nature of expression and response as they spontaneously enter the actor's mind. Enacting spontaneity requires the actor to suppress the desire to dictate the terms of an exchange, as Johnstone (1981: 32) explains:

> I'm teaching spontaneity, and therefore I tell them that they mustn't try and control the future, or to 'win'; and that they're to have an empty head and just watch. When it's their turn to take part they're to come out and just do what they're asked to, and *see what happens.*

This account presumes that an actor who has succeeded in presenting a 'true' account of improvisation has suppressed all other calculations and presented a pure form of free will. This glimpse of a pre-cognitive essence of human expression has seen improvisation loaded with positive connotation. As Frost and Yarrow explain (1989: 3), this theatrical framework is suggestive of some quest for truth:

> [...] for improvisation is about order, and about adaption, and about truthfully responding to changing circumstances, and about generating meaning out of contextual accidents.

There are a number of implications of understanding improvisation in these virtuous terms. First, and perhaps most profoundly, scholars often assume improvisation to denote some form of subversive action that relates to the utterance of truths that may destabilize existing relations of power. This account creates a vertical hierarchy, where the performer may challenge those 'above' by refusing or adapting existing scripts. Such an imagined hierarchy is granted some credence by state attempts to set limits to the public performance of improvisation. For example, in the UK the 1848 Theatres Act demanded that the Lord Chamberlain grant permission prior to the performance of new theatrical productions, often censoring parts that were considered against public decency or politically subversive. Consequently, until the Act was repealed in 1968 the public performance of improvisation was illegal in the UK. The second, and related, aspect of a virtuous reading of this theatrical understanding of improvisation is that it places emphasis on the disposition of the improviser as a knowing subject. Both Johnstone (1981) and Frost and Yarrow (1989) emphasize the individual preparation required to improvise: to be alert, prepared and ready to respond. This is suggestive of a form of free agency where the individual has the capacity to use their imagination to improvise previously unseen performances. This understanding of the term recalls Johnstone's (1981) point that improvisation may be learnt, that through training the individual may perfect the forms of mental preparation required to enact an improvisation.

But we need to be cautious in transposing these understandings of improvisation from theatre craft to the social sciences. Such virtuous understandings of the term are reliant on an imagined pure free will which is eroded by formal education and enculturation. Within this optic the successful practice of improvisation stands as a moment outside these other constraints and preoccupations. But this theatrical ideal rests on the possibility of bracketing all forms of value judgement and hierarchy from the moment of improvisation. For example, when Johnstone speaks of the corrosive effect of formal education in denying the 'real self' he sets his form of dramatic training apart as an authentic attempt to strip back layers of enculturation. But of course, the structure of Johnstone's tuition needs to be considered as another form of conditioning that shapes the perception and performance of the subsequent improvisations. We could read Johnstone's attempts to uncover 'real' spontaneity less as the revelation of the natural essence of the performer and more in terms of an altering calculation by the performer as to what is appropriate to the relative value of different performances. In this case the actor has transformed their calculation of what is considered appropriate behaviour from one form of expression (say, being funny or trying to be original) to another form of expression (the impression of an empty head) that is stated to be more valued by Johnstone. The underlying practice of improvisation remains resourcefulness, despite the altered nature of the physical expression of their performance.

The tension between imagined spontaneity and underlying resourcefulness is expressed in scholarly applications of improvisation. In Philip Crang's

(1994: 681) account of waiting tables he talks of using improvisation to transgress pre-established routines, practices and identities:

> [...] experiences of doing the work were less dominated by a reduction of the self on display to a prewritten script, and revolved around rather more improvisational performances; performances that used and at times abused a script such as the order of service, but did not simply enact it; performances that involved social relations of display that were more than the projections of managerial surveillance strategies.

In this case improvisation is a form of individual adjustment of pre-given scripts in order to suit the needs of a particular context. Improvisation stands as a means of adaptation. But Crang is clear that these are not simply spur-of-the-moment iterations, but are rather prefigured and constrained by wider structures that shape his action. While his performances operated outwith the surveillance strategies of the restaurant management, they were shaped by other concerns. Crang is clear that much of his activity in the restaurant was a calculated attempt to increase his tips from customers. While his performances may have deviated from the specific text of preordained scripts, they continued to reproduce management demands for a creative and entertaining waiter. Rather than being spontaneous, Crang is using the term improvisation as a means of conveying 'resourcefulness' within a context of limited possibilities. Thus, even when operating 'on the spur of the moment' the decisions made reflect wider ethical, political or economic considerations.

The term improvisation, then, highlights moments of the admixture of creativity and production, rather than instances of unbridled free agency. Indeed, the notion of improvisation holds in tension structure and agency, illustrating that these are negotiated through practice and cannot be rendered in purely abstract terms. By focusing on a sense of 'performed resourcefulness' this interpretation of improvisation is indebted to the structural anthropology of Claude Lévi-Strauss, in particular *The Savage Mind* (1972). Lévi-Strauss used the term *bricolage* to intimate the way in which non-Western societies make sense of the world through 'making do' with available social categories and symbols (see also Hebdige, 1979). This approach is rooted in a syntactic understanding of social forms as related through grammar relations, where society fits together 'like words in a sentence, to form a meaningful whole' (Garvey, 1971: 11). Through this framework Lévi-Strauss presents the figure of the *bricoleur*, one who makes use of what is available to produce something new, as opposed to the engineer who designs solutions that specify requirements for particular tools, skills and materials (Baker, 2007: 697). The bricoleur's resourcefulness takes material and ideational forms. In material terms Lévi-Strauss talks of a 'science of the concrete' where the bricoleur has a deeper knowledge of the fabric of the environment which allows them to take advantage of what is cheaply or freely available. In contrast, ideational bricolage refers to the practice of drawing on older myths and beliefs to serve new symbolic purposes. A good example of such ideational bricolage has been provided by the recent work of scholars of nationalism, who illustrate how newly formed nationalist movements draw on symbols of the past to bolster emerging political projects (see for example Dixon, 2008).

While bricolage illuminates the significance of the skills set of the individual and the centrality of surrounding resources, there are dangers in conflating bricolage and improvisation. Specifically, scholars have suggested that *The Savage Mind* offers a structural account of social ordering which underplays forms of domination that shape processes of bricolage (Werbner, 1986). There is a risk that Lévi-Strauss's approach renders historical narratives of exploitation in terms of differing essences of human capacity. Most specifically, Lévi-Strauss's distinction between the engineered nature of societies within the Global North and the bricolage of the Global South risks reproducing colonial binaries of civilized and traditional populations. In some senses this critique points to the inherent limitations of Lévi-Strauss's macrosociological approach, which was less attuned to the existential questions of the nature of subjectivity and more concerned with broader notions of cultural traits.

In contrast to this structuralist approach, Pierre Bourdieu (1984, 1987, 1990) sought to present a theory of social practice which was more attentive to both uneven geographies of power and the forms of human agency that this cultivates. It is work that provides a more balanced assessment of the structure/agency divide, providing the vocabulary to think through the interplay of powerful agencies with individual subjects. As Jeffrey *et al.*'s (2008) work on education in north India notes, Bourdieu provides a set of conceptual tools to understand both the mechanisms used by individuals to improve their status and the range of structural factors that guide their strategies. One of the enduring metaphors used by Bourdieu is to equate the social world with a game, where strategies are the product of conscious and rational calculation within a bound set of rules. But the game's conventions alone do not govern individual behaviour, since to excel at the game requires 'constant innovation and improvisation which goes well beyond the simple following of rules and regulations' (Cresswell, 2002: 381). In order to convey this rather abstract knowledge of the appropriate course of action, Bourdieu points to the significance of 'practical sense', where the playing of the games becomes part of nature (Cresswell, 2002).

In order to trace the accomplishment of 'practical sense' in social life, Bourdieu sought to explore the accumulation of different varieties of capital. For Bourdieu, competition over capital highlights 'the struggle for scarce goods for which the universe is the site' (Bourdieu, 1987: 3). He continues:

> [The] [...] fundamental social powers are, according to my empirical observations, firstly *economic capital*, in its various kinds; secondly *cultural capital* [...], again in its various kinds and thirdly two forms of capital that are very strongly correlated, *social capital*, which consists of resources based on connections and group membership, and *symbolic capital*, which is the form the different forms of capital take once they are perceived and recognized as legitimate. (Bourdieu, 1987: 3–4, original emphasis)

One of Bourdieu's key insights was that each form of capital could be converted into other forms (Painter, 2000: 244), though with varying degrees of difficulty (Calhoun, 1993: 69). Bourdieu conceived symbolic capital as the most influential as 'it is the power granted to those who have obtained sufficient recognition to be in a position to impose recognition' (Bourdieu, 1989: 23). Crucially, a key aspect of this power is control over other forms of capital, managing their conversion and exchange rates (Engler, 2003: 449). As discussed in Chapter Four, this type of capital has been referred to as 'statist capital', as it is the state that has the legitimacy to confer value on certain activities, while devaluing others (Engler, 2003). Viewing social and political practices in terms of capital privileges neither structure nor agency; individuals can exercise their agency, but are still constrained (or enabled) by their reserves of social, cultural and economic capital (Bridge, 2001).

The structure/agency binary permeates the work of Bourdieu as he attempts to construct a theory of practice that overcomes the theoretical oppositions between binaries such as subjectivism and objectivism; particularism and universalism; and agency and structure. Central to this objective are Bourdieu's notions of *habitus* and *field*. Bourdieu describes *habitus* as 'a system of durable, transposable dispositions which functions as the generative basis of structured, objectively unified practices' (Bourdieu, 1979: 11). For Painter (2000) this makes *habitus* a 'mediating link between objective social structures and individual action and refers to the embodiment in individual actors of a system of norms, understandings and patterns of behaviour, which, while not wholly determining action [...] do ensure that individuals are more disposed to act in some ways rather than others' (Painter, 2000: 242). What appears critical is that *habitus* functions *below* the level of consciousness and language, 'beyond the reach of introspective scrutiny or control by will' (Bourdieu, 1984: 466). In this way, *habitus* has been understood as both 'the social inscribed on the body of the individual' (Cresswell, 2002: 381) and 'the regulated source of improvisations' (Calhoun, 1993: 78). In addition, Engler (2003) points to the reflexivity of *habitus* as it is both 'structured by, and structuring, each individual social position' (Engler, 2003: 450).

In contrast to the embodied and tacit nature of *habitus*, Bourdieu's notion of *field* is a more grounded conceptual tool. Bourdieu sees the social world as divided into a series of *fields*, each of which constitutes a site of a struggle over a particular form of capital (Harker, 1990: 97). As Painter (2000) points out, this concept mediates Bourdieu's other theoretical devices as individual actors bring to the *field* both the embodied disposition of the *habitus* and their stocks of accumulated capital (Painter, 2000: 245). As with *habitus*, Bourdieu was keen to demonstrate that each *field* was dynamic, as struggles are not only over forms of capital but also over the power to define the *field* itself (Postone *et al.*, 1993: 6). The actions of individuals within the *field* are guided by 'strategy', which again breaks with subjectivist and objectivist thinking in viewing decision making as neither conscious, nor calculated, nor mechanically determined (Mahar *et al.*, 1990: 17). Instead, as discussed above, strategy is the intuitive product of 'a practical sense of a particular social game' (Cresswell, 2002: 380). This rhetoric of 'feel for the game' has drawn similarities in some quarters between Bourdieu's notion of strategy and rational actor theory (RAT) (Bridge, 2001: 209; Painter, 2000: 245). Bourdieu, however, rejects RAT as part of a wider aversion to subjectivism (Bourdieu and Wacquant, 1992: 118). RAT is dismissed as an overly economistic approach to life, without an appreciation of how actors are disposed to make certain decisions to maximize capital while guided by the logic of the *habitus*. This schema has particular bearings on Bourdieu's understanding of the state. For Bourdieu the technologies and tactics through which the state reproduced its power were only successful on account of the state's accumulation of symbolic capital (Bourdieu, 1984, 1989). One attribute of such capital is, according to Bourdieu (1989), that it confers upon the perspective of the state an absolute, universal value, thus 'snatching it from a relativity that is by definition inherent in every point of view, as a view taken from a particular point in social space' (Bourdieu, 1989: 22). Such mechanisms of legitimization can lead to what Bourdieu refers to as *symbolic violence*, as those who do not have the 'means of speech' or do not know how to 'take the floor' can only see themselves in the words or the discourses of others – that is, those who can name and represent (Mahar *et al.*, 1990: 14).

2.4   Improvising the State

This final point is central to the idea of the improvised state. Bringing together improvisation and the state is, at first glance, an attempt to integrate the ephemeral (improvisation) with the durable (the state). But Bourdieu's conceptual vocabulary unsettles both these characterizations, as improvisation becomes more durable (through an analysis of the *habitus* and capital which shape state decisions) and the state more ephemeral (as the achievement of statehood is explored through the practice of accumulating and deploying symbolic capital). One of the enduring refrains of this book is the way in which statecraft relies on fleeting moments of performed authority that are enacted 'as if' they are based on timeless claims to power. The fragile achievement of BiH as a state relies on the repeated performance of its existence, including material cultures of currencies and flags, through to spatial practices of securing borders. But one of the central points is that there is not a unitary vision of statehood in BiH; rather, plural claims to symbolic capital exist simultaneously and seek to cultivate their legitimacy through a range of alternative material and symbolic practices.

By understanding the state as improvised we may begin to explore the dialectic nature of structure and agency within contemporary BiH, where actors are neither engaging in performances of pure free will nor beholden to rigid economic or political structures. In this respect the argument follows a long lineage of post-Marxist attempts to transcend this binary, from structuration theory (Giddens, 1984) through to strategic-relational approaches (Jessop, 1990, 2005, 2008; Jessop et al., 2008). In the latter work, Bob Jessop comes closest to encapsulating the conception of the improvised state by arguing that the 'emergence of relatively stable structural ensembles involves not only the conduct of agents and their conditions of action but also the very constitution of agents, identities, interests, and strategies' (Jessop, 2005: 53). This perspective reflects Jessop's close reading of Poulantzas (1978; see also section 2.1 above) in charting the specific material and historical grounding of the co-emergence of structure and agency within the experience of specific states. Consequently, Jessop encourages exploration of the spatio-temporal context of state projects and critiques a purely discursive approach that fails to grasp the limits to the agency of state practices (see Jessop, 2001). As the name would suggest, the strategic-relational approach orientates attention to the nature and evolution of strategic elements of state practice, foregrounding in particular how the changing nature of the state reflects changes in capitalist systems of production. It also brings to the fore relationality, that the state is a socially embedded phenomenon that cannot be considered as a distinct actor divorced from a wider society.

While sharing much in terms of theoretical departure points, perhaps the clearest overlap across the strategic-relational approach and the lens of improvisation is the attempt to capture what Jessop et al. (2008: 390) term 'polymorphy', where the organization of socio-spatial relations is analysed across 'multiple forms and dimensions'. Hence, the concept of improvisation is utilized in order to avoid privileging a single framework of spatial analysis (such as territory, scale, network, place or space) and instead explores the ways in which different geographical referents have been used to legitimize specific (and often competing) state projects. In addition, I share with Jessop the desire to advocate improvisation as a means of analysis, rather than a single unified theory of the state. The key to the strategic-relational approach is its advocacy of understanding the spatial and temporary specificity of socio-spatial relationships. In terms of the development of state theory, this has equated to Jessop's long-term engagement with the development of the capitalist state in the UK, from the crisis of Fordism following the Second World War to the Keynesianism of the welfare state, through to Thatcherism and beyond (see Jessop, 2008: 23).

But improvisation should not be taken as a subset of the strategic-relational approach. The key contribution of improvisation is the attempt to disperse state theory through a range of settings, relationships and dispositions. While the implications of grand historical shifts in BiH are clear (and explored in the following chapter and beyond), the lens of improvisation seeks to examine how different interpretations of history and geography are performed in the present, and in doing so to advocate different ideas of the present and future state. This fragments state theory, rather than providing the integrative framework of either the strategic-relational approach or structuration theory. What is left is a sense of the limits to pure theoretical reflection in the absence of empirical engagement and experience. In this sense, improvisation seeks to illuminate the very unfolding of ideas of the state as they are articulated and practised across a range of socio-spatial settings, many of which are not the classic domain of state bureaucrats, policy makers or politicians.

In this book I will take forward Bourdieu's conceptual vocabulary to understand how these competing ideas of the state circulate and are resisted following the signing of the Dayton GFAP in 1995. This approach allows the book to focus on the historical and spatial contexts within which ideas of the state are pursued. Specifically, it allows us to examine the resources available to both political actors and civil society organizations, forms of capital that can both cultivate and destabilize a sense of state legitimacy for both domestic and international audiences. One of the key resources for aspiring state elites has been geopolitical narratives of the past. Therefore, in order to understand how and why particular improvisations of the state have become naturalized and reproduced, we need to investigate the competing understandings of history.

Chapter Three

Producing Bosnia and Herzegovina

[E]ach people was perpetually making charges of inhumanity against all its neighbours. The Serb, for example, raised his [sic] bitterest complaints against the Turk, but was also ready to accuse the Greeks, the Bulgarians, the Vlachs, and the Albanians for every crime under the sun. English persons, therefore, of humanitarian and reformist disposition constantly went out to the Balkan Peninsula to see who was in fact ill-treating whom, and, being by the very nature of their perfectionist faith unable to accept the horrid hypothesis that everybody was ill-treating everybody else, all came back with a pet Balkan people established in their hearts as suffering and innocent, eternally the massacree and never the massacrer. The same sort of person, devoted to good works and austerities, who is traditionally supposed to keep a cat and a parrot, often set up on the hearth the image of the Albanian or the Bulgarian or the Serbian or the Macedonian Greek people, which had all the force and blandness of pious fantasy.

West (2001 [1941]: 20)

One of the primary findings of critical scholarship on the origins of the conflict in BiH between 1992 and 1995 has been the futility of seeking to establish a single cause of the conflict (Campbell, 1998a, 1999; Simms, 2001). Rebecca West's notorious remark that scholars return from travels to the then Yugoslav region with a 'pet Balkan people' is at first sight a useful statement of the subjectivity of researchers and journalists and the partial histories they write. But her words are also an early indication of the rigid categorizations that shape accounts of the past and present in the former Yugoslavia. The reduction of political life to simple 'ethnic' or 'national'

The Improvised State: Sovereignty, Performance and Agency in Dayton Bosnia, First Edition. Alex Jeffrey.
© 2013 John Wiley & Sons, Ltd. Published 2013 by John Wiley & Sons, Ltd.

groupings acknowledges the centrality of these terms within political discourse, but overstates their significance to social or political life. Bose (2002: 10) is right to suggest that such simple renderings 'do enormous violence to complex Bosnian realities', as numerous anthropological accounts written prior to and since the Bosnian war have illustrated (Bringa, 1995; Jansen, 2007; Jašarević, 2007). The author of a history of BiH therefore faces the dual task of emphasizing the prominence of such discourse while challenging its ontological basis. This approach requires a shift away from an unproblematic vocabulary of politics structured around imagined identity labels and transhistorical geographical referents, towards an understanding of the forms of social and spatial practices that sustain such labels and affiliated claims to state sovereignty.

Such a critical rendering of history builds on the previous chapter's theorization of the state as performed resourcefulness. Specifically, the multiple coexisting performances of BiH following the Dayton Agreement are rooted in competing understandings of the past. In order to lay claim to legitimacy and command respect, policy makers and political actors have grounded their claims in different historical narratives with competing geopolitical underpinnings. The lack of consensus about the history of the former Yugoslavia is well understood (see Campbell, 1998b), and the divisive nature of parallel historical narratives of recent events has been well documented (Robinson *et al.*, 2001). Charles Ingrao and Thomas Emmert (2009) have attempted to confront these challenges through the creation of a single historical narrative of the fragmentation of Yugoslavia written by a team of scholars from both the Yugoslav region and beyond. While the book itself successfully constructs a shared account of the key events in the destruction of the Yugoslav state, its critical reception (both in the former Yugoslavia and in the United States) points to the ongoing political divisions concerning differing renditions of the past (Ingrao, 2011). As debate surrounding Ingrao and Emmert's book attests, the writing of history is not a neutral and descriptive exercise; it is a practice that enters into the constitution of the past and lends legitimacy to certain conceptualizations of the present (Campbell, 1998b). Consequently, the terms and vocabulary used to understand the rise and fall of Yugoslavia shape what can be constituted as a just outcome to the violence.

This chapter explores how scholars have accounted for the differing performances of state sovereignty in the territory of first Yugoslavia and then later BiH. This is not a conventional historical account of the end of empire on the Balkan Peninsula and the rise and subsequent fall of Yugoslavia.[1] This chapter has more modest ambitions. Through an examination of specific historical events, the analysis will focus on plural geopolitical practices that have been enacted over the territory of Yugoslavia. This centres the analysis on questions of space: how abstract ideas of solidarity and community have been justified through practices of bordering, exclusion and state building. Engaging with such a history of Yugoslavia requires understanding the deployment of geographical language to claim the inevitability of certain political outcomes (Campbell, 1998a; Kearns, 2009; Kuus, 2007). As discussed in Chapter One, scholars of critical geopolitics have examined the histories and practices of geopolitics as discursive exercises through which political relationships are ordered, understood and, crucially, spatialized (see Ó Tuathail and Agnew, 1992: 190). But as Kuus (2007) has argued, such insights are more than an argument for 'adding geography in' to the study of international relations. Instead, the discursive approach advanced by critical scholars draws attention to the mechanisms through which geopolitical language 'naturalises and normalises particular political claims as "geographical" and hence given' (Kuus, 2007: 7).

A critical perspective on the history of conflict in BiH demands deconstructing the spatial registers used to frame the fragmentation of Yugoslavia and the establishment of successor states. This approach is indebted to narrativist understandings of historiography, approaches that have illustrated the act of writing history as an exercise in the construction of meaning (White, 1973). In this chapter, three specific spatial narratives are examined, and in doing so some of the competing understandings of the violence between 1992 and 1995 are brought to the fore. The first section explores the framing of BiH as a 'fault line' between imperial powers. The location of BiH as variously within and between the Holy Roman, Ottoman and Austro-Hungarian empires has been used by historians and politicians as explanation for the emergence of violence across the twentieth century. The second section examines the presentation of BiH as part of a 'barrier' between East and West. The peace settlements at the end of the First and Second World Wars saw the establishment of a strong Yugoslav state as a security cordon between Western Europe and the Soviet Union. This geopolitical register ensured strong support from Western European powers for political elites who sought to retain the territorial integrity of the Yugoslav state. While the first two sections examine liminal spaces at the boundaries of other polities, BiH is located at the centre of the third spatial register under consideration. Artistic and literary representations of BiH over the past two hundred years have placed this territory at the heart of an imagined 'Balkans'. But this representation is not a static backdrop. Representations of the Balkans are often structured around a sense of the attraction of this region to those outside, often using the metaphor of the 'vortex' or 'whirlpool'. This geography has had profound political implications, in particular during the fragmentation of Yugoslavia where political leaders often presented the violence as a natural and inevitable expression of BiH's Balkan heritage, which posed a risk to regional powers.

Reflecting a characteristic of classical geopolitical narratives (see Kearns, 2009), each of the spatial narratives covered in this chapter has, to varying degrees, sought to present political confrontation as an inevitability, grounded in the geography of the region. The three sections do not follow a strict chronology, since the geographical imaginaries covered in each section have been conveyed at different times through history as political elites have attempted to justify particular configurations of the present. The final section of the chapter examines the forms of compromise fostered at the GFAP, which constitutes one of the most tangible attempts to accommodate these divergent spatial histories. The negotiators at Dayton improvised the post-conflict BiH state by drawing on the symbolic resources provided by these three narratives of history. What transpired was a form of political compromise that deploys traits of all three: it partitioned BiH space into ethnic spaces (fault line) and retained a BiH state (barrier) while seeking to limit (initially at least) regional and wider international involvement (vortex).

3.1   Fault Line

The field of tectonic geomorphology has provided a repository of characteristics used by scholars of international relations to describe world affairs. Samuel Huntington's (1997) thesis of a 'clash of civilizations' is perhaps the most prominent example of such geopolitical reasoning, where the author suggested that the world could be divided into eight civilizations, divisions which would shape the nature and location of future global conflicts. Huntington argues that conflict may be traced to the perimeters of these civilizations where adjacent cultural differences act as the catalyst for violent conflict (see Bieber, 1999; Hansen, 2006). More recently Thomas Barnett (2005) has attempted to illustrate new security concerns facing the United States in the wake of 9/11 by tracing similar global dynamics. While the most eye-catching part of Barnett's geography may be his assertion that the world could be divided neatly into an 'integrated core' and a 'non-integrating gap', his thesis also suggests that it was along the line between these two territorial entities that conflicts may be predicted (though as is the case in seismology, without any precise timings of violent activity). Barnett's imagery is visually reminiscent of tectonic mapping, with the two areas presented as plates of the earth's crust along which violence could erupt. The reasoning of both Huntington and Barnett is reminiscent of classical geopolitical thinking: a global vision of world politics that seeks to identify a unifying narrative to explain an array of tensions and contestations. This approach does not simply erase geography, as Ó Tuathail (1996) argued in the case of Halford Mackinder's classical geopolitical thinking. Rather than engaging with the multi-scalar set of forces that shape the outbreak of specific conflicts, this style of 'fault line' reasoning localizes violence as a consequence of abrasions between core and gap. For Huntington and Barnett, being located at the fault line explains the onset of violence.

The image of the 'fault line' has been prevalent in accounts of the fragmentation of Yugoslavia and, in particular, the war in BiH. There are a series of historical events that support the narrative of civilizational 'fault line'. Historians have identified the death of Emperor Theodosius I and the subsequent east–west division of the Roman Empire between his two sons Arcadius and Honorius in AD 395 as a key moment in the establishment of a civilizational cleavage in southeast Europe (Gibbon, 1828: 114). In the abstract this vision has the potential to project a conception of modern state territorialization and boundaries back into an era of radically different forms and hierarchies of geopolitical rule, in particular if this territorial division is expressed cartographically. But, as Tim Judah (2000) has noted, the impact of the division of the Roman Empire was most tangible in the field of religious emissaries and influence. For example, around the fifth century AD the territories that roughly cover present-day Serbia, Montenegro and the Former Yugoslav Republic of Macedonia (FYROM) increasingly hosted missionaries from Orthodox Byzantium, while present-day Slovenia and Croatia remained in the sphere of influence of Catholic Rome (see Judah, 2000: 8–10). The split between these two powers was not, then, a neat boundary but a mobile and diffused frontier that shifted through the subsequent centuries, in particular during the Justinian Dynasty of the Byzantine Empire in the sixth and seventh centuries. Despite this mobility, the divided religious loyalties went on to influence wider cultural practices, principally through the role of missionaries in shaping the production of knowledge concerning social and cultural life. Most notably, in the eighth century St Cyril, a Greek Orthodox missionary to the Serbs, translated the Gospels and Orthodox liturgy into Serbo-Croatian, thus establishing the Cyrillic alphabet. This move created what has become a key symbolic difference between Slav peoples across the territory of the former Yugoslavia: while Croat, Bosnian-Muslim and Slovenian groups use the Latin alphabet, those identifying as Serb often use Cyrillic.

The origins of BiH's 'fault line' analogy challenge the precision of this boundary, suggesting instead a frontier region which saw comparatively frequent shifts in the spatial organization of authority. Significantly, the centuries following the schism in the Roman Empire witnessed rising antagonism between two expansionist political entities: an emergent Serbian Kingdom (later referred to by some as the Serbian Empire) and the Ottoman Empire. The Serbian Kingdom was established in the eleventh century by Stefan Nemanja, who became the head of a dynasty that lasted over two hundred years. Historians have identified two aspects to Nemanja's initial success in establishing a distinct Serb territory. The first was his diplomatic and military aptitude, whereby he forged crucial allegiances with Bulgars and Magyars in order to challenge by military means the dominance of Byzantine rule. Second, he strengthened the internal cohesion of the kingdom through the introduction of the Serb Orthodox faith as the single religion and institutionalized education using the Cyrillic script (Hajek, 1993). The kingdom reached its political and territorial peak in the early fourteenth century under the stewardship of Stephen Uroš IV Dusan of Serbia, covering the territory of present-day Serbia, FYROM and northern parts of Greece. However, historians note that in the absence of a universally accepted successor, the death of Uroš IV marked the start of the kingdom's decline and disintegration (Dvornik, 1962).

At this time, the power of the Byzantine Empire on the Anatolian peninsula was waning and its territory was divided between prominent ruling warriors (*Ghazi*). As Byzantine power shrank, so the Ghazi looked west to find areas to expand, leading Tamim Ansary to suggest that the frontier marches were 'mother's milk to the Ghazi emirates' (Ansary, 2009: 175). One of the leading Ghazi emirates of the late thirteenth century was ruled by Othman (also known as Osman), a descendant of pastoral nomads from Central Asia and a skilled horseman and fighter. Historical records of this central figure in the establishment of the Ottoman Empire are sparse and predominantly rely upon medieval chroniclers, who often embellished accounts with extravagant tales of heroism or religious virtue (see Kermeli, 2008). However, Ansary (2009: 172) paints a picture of this young Ghazi using his combat skills to his financial and territorial advantage:

> In the fighting season he would lead his men to the frontier provinces and accumulate booty by fighting Christians. In the 'off-season,' he collected taxes from any productive settled folks he found in the areas he controlled.

It is significant that Ansary suggests that it was this *process* of plunder and tribute rather than a territory or empire that was the most significant bequest by Othman to his descendants. Othman was improvising his rule, performing his power through military supremacy and fiscal control. Medieval chronicles view the defeat of the Byzantine army at Baphus in 1301 as a key moment in the emergence of Othman's (Ottoman) empire. By the time of Othman's death in 1324 he had established a considerable sphere of influence in southeast Europe by annexing Ghazi emirates and continuing to stage incursions westwards into the Balkan Peninsula. Othman was succeeded by his two sons, first Orhan I (ruling from 1324 to 1362) and subsequently Murad I (ruling from 1362 to 1389).

These portraits of the Serbian Kingdom and fledgling Ottoman Empire illustrate two dispersed sets of spatial practices that relied on forging senses of solidarity through a range of social, military and fiscal processes. The concept of territory within these realms does not conform to modern notions of state territoriality involving a neat correspondence between territory and sovereignty (see Elden, forthcoming). These jurisdictions are best understood as competing spheres of influence rather than precise territorializations that could claim homogeneous or exclusive populations. But this narrative of spatial and social heterogeneity has been lost in the subsequent political interpretations of the relationship between the Serbian Kingdom and the Ottoman Empire. The antagonism between these two ancient polities has provided a rich set of reference points to justify political cleavages in the post-Cold War world. Most notably, the battles between forces loyal to the Serbian Kingdom and those hailing from the Ottoman Empire in the fourteenth century have assumed particular symbolism within narratives of Serb nationalism and fault line geopolitics, founded as these both are on essentialized notions of identity and territory.

The symbolic import of Serbian and Ottoman antagonism is most evident in the present-day political deployments of Serb victimhood at the Battle of Kosovo in 1389. Indeed, it is challenging to analyse the nature of this battle in isolation from the powerful discourse of heroic Serbian sacrifice that has framed some subsequent commentaries (for a critique, see Mertus, 1999). However, historians have placed the Battle of Kosovo into a context of a series of skirmishes between the Serb and Ottoman forces, in particular noting that it followed a battle at River Maritsa in 1371 in present-day Bulgaria. Crucially, these accounts view the battle at River Maritsa as being of greater significance to the geopolitics of the region, since it marked a comprehensive defeat for Serb forces and a critical moment of reversal for the expansionist Serbian Kingdom established by Stefan Nemanja (see Nolan, 2006: 575). Despite the outcome of the 1371 battle, it was the encounter eighteen years later at Kosovo Polje that has assumed mythical status within Serb national consciousness. Radha Kumar (1997: 3) suggests that the immediate significance of the Battle of Kosovo in 1389 was negligible and it is best seen as 'more of a draw' than a heroic defeat for the Serbs marking the beginning of several centuries under Ottoman rule. Misha Glenny (1999) points out that Serbian power collapsed gradually over the following sixty years and '[t]he fortress in Belgrade did not fall under Ottoman control until the sixteenth century' (Glenny, 1999: 11). However, the battle did mark the deaths of leaders on both sides: Ottoman Sultan Murad I was killed by Miloš Obilić (a Serb knight who is reported to have used a ruse to gain access to, and subsequently stab, the Ottoman ruler), while Serb Prince Lazar died in the course of the fighting (Anzulović, 1999: 13–14).

The subsequent interpretation of the Battle of Kosovo has amplified its political significance. Two sets of sources are available that recount certain events of the battle while also producing (and reproducing) certain narratives about Serbian identity, heroism and destiny. The first is the chronicles of the battle and its aftermath produced by Serb Orthodox patriarchs. In a survey of these documents the historian Thomas A. Emmert (1990: 68) has identified a series of key schisms:

> The writers of the battle very simply describe the battle as a struggle between the forces of good and evil: Murad with his band of bloodthirsty beasts and Lazar with his pious army of God-fearing Christians. The Turks are identified as Ishmaelites or Hagarites – an obvious and derisive reference to the Old Testament. In *Zitije kneza Lazara* [The Life of Prince Lazar, an anonymous medieval Serb chronicle] the Ishmaelites are 'arrows released by God because of our sins' and Murad is 'the beast who came roaring like a lion to devour Christ's flock and destroy our homeland.'

One of the most enduring narratives that emerges from the chronicles surrounds the apparent choice made by Prince Lazar at his death. *Slovo o knezu Lazaru* [Narration about Prince Lazar], a fourteenth-century chronicle written by the Serb Patriach Danilo III, suggests that Lazar was a martyr who chose an eternal 'kingdom in heaven' over a temporal 'kingdom on earth' (see Anzulović, 1999: 11). This imagery asserts political agency into this supposed defeat, where the sacrifice of the individual established a form of sacred permanence.

The images and narratives of the chronicles have fed into a second set of sources: Serb epic poetry. This highly developed oral folk tradition comprises poems relating to events and individuals from Serbian history and the majority were composed between the fourteenth and nineteenth centuries. The first major written collection was compiled in the early nineteenth century by the Serbian linguist Vuk Karadžić (1787–1864). Recitations of the poetry are often accompanied by the gusle, a single-stringed instrument played with a bow. In the preface to a 1980s edition of epic poetry, the poet Charles Simić (in Matthias and Vučković, 1987: 8) paints a vivid picture of listening to a rendition of epic poetry while at school in Belgrade in the late 1940s:

> I still remember my astonishment at what I heard. I suppose I expected the old instrument to sound beautiful, the singing to be inspiring as our history books told us was the case. *Gusle*, however, can hardly be heard in a large room. The sound of that one string is faint, rasping, screechy, tentative. The chanting that goes with it is toneless, monotonous, and unrelieved by vocal flourishes of any kind. The singer simply doesn't show off. There's nothing to do but pay close attention to the words which the *guslar* enunciates with great emphasis and clarity. [...] After a while, the poem and the archaic, other-worldly-sounding instrument began to get to me and everybody else. Our anonymous ancestor poet knew what he was doing. This stubborn drone combined with the sublime lyricism of the poem touched the rawest spot in our psyche. The old wounds were reopened.

It is Simić's allusion to the affective power of Serb poetry to collapse the distance between past and present that speaks to their significance as political instruments. While the texts of Serb epic poetry often present nuanced characterizations that challenge the heroism and victim narratives seen in the religious chronicles, it is through their performance that a sense of a transhistorical Serbian identity that faces perpetual existential threats may be conveyed, not least from an imagined (and equally timeless) Ottoman Empire. Geoffrey Locke (2002: xvii), in an exuberant foreword to an English language edition of Serbian epic poetry, points to the key facets of the Battle of Kosovo myth conveyed through these verses:

> [...] the Serbian perception was that it was at Kosovo that they had thrown everything they had into the balance, and that it was there that the flower of their chivalry and manhood had sacrificed itself in a final effort to defend the nation from slavery and Europe from Islamic domination. [...] [T]hey saw that before Kosovo they had been an independent and justifiably proud kingdom: after it nothing could halt their decline into vassaldom under a foreign and heathen occupier, stripped of their wealth and dignity, but never their pride.

We see through the chronicles and epic poems the reification of certain roles, images and relationships which have become established political devices both in Yugoslav politics and further afield. Here the image of the fault line finds commodification in tales and verses, forms of symbolism that shape public commemoration and social life. Historians and political commentators have noted the way in which references to civilizational struggles have been deployed by political agents in order to legitimize particular agendas, most notably citing Slobodan Milošević's speech at the 600th anniversary of the Battle of Kosovo on 28 June 1989 (see Judah, 2000). Coming at key moments in both the fragmentation of Yugoslavia and Milošević's rise to power, the Gazimestan Speech (as it is known) certainly deployed imagery from the Battle of Kosovo in order to bolster senses of Serbian victimhood, citing in particular an historic fault line:

> What has been certain through all the centuries until our time today is that disharmony struck Kosovo 600 years ago. If we lost the battle, then this was not only the result of social superiority and the armed advantage of the Ottoman Empire but also of the tragic disunity in the leadership of the Serbian state at that time. In that distant 1389, the Ottoman Empire was not only stronger than that of the Serbs but it was also more fortunate than the Serbian kingdom. (Cited in Krieger, 2001: 10)

What seems crucial to this account is the centrality of spatiality in determining difference, a form of geopolitical reasoning from which we can discern two interlinked implications. First, as detailed in the introduction to the chapter, the enrolment of geography emphasizes a sense that political or cultural differences are 'natural' and stem from divisions in the earth's crust. Antagonism and violence thereby may be perceived as externalities of a natural process, rather than human practices enacted to bolster individual or collective agendas. Second, the commodification of the fault line in story and verse allows this geographical referent to conjure a series of cultural, political and religious significations. More than simply designating 'self' and 'other', the image of the fault line has been used politically to refer to a complex collection of narratives encircling sacrifice, martyrdom and hallowed soil. Entangling the sacred and the profane, these scripts have allowed the image of the fault line to be used by political agents within and beyond the borders of the former Yugoslavia as a flexible signifier of simultaneous loss, moral courage and entitlement.

As Merje Kuus (2007: 10) has illustrated, such geopolitical slogans do not cause particular policy outcomes, but they rather frame political debate 'in such a way as to make certain policies seem reasonable and feasible while marginalizing other policy options as unreasonable and unfeasible'. As we will see, the image of BiH on a geopolitical fault line shaped many of the interpretations of the violence of the 1990s and the subsequent forms of state sovereignty that emerged within the region. Reliance on this narrative suggested that fault lines needed to be formalized through political boundaries rather than transcended through alternative understandings of political life. This sense of political geography reflecting 'nature' sets in place a normative commitment behind subsequent interventions. The central political consequence of fault line geopolitics is to cultivate a sense of inevitability: that certain outcomes in the fragmentation of Yugoslavia were predetermined and subsequent interventions should mirror such ancient spatial cleavages.

3.2   Barrier

> The interests of all the peoples making up Yugoslavia are so closely related and their neighbors are so formidable in size and strength that a confederated union, economic as well as political, is required.
>
> Bowman (1928: 355)

The fault line analogy presents the territory of the former Yugsolavia as a site of abrasion and conflict. But this spatial analogy exists alongside other, often contradictory, geopolitical scripts. Perhaps the most prevalent such script presents Yugoslavia as a 'buffer' or 'barrier' *between* territories. One of the most widely cited reasons given by historians for the fragmentation of Yugoslavia is the loss of its role within the Cold War as a 'buffer zone' between capitalist West and socialist East (Papandreou, 2000: 162; Wachtel and Bennett, 2009: 42). As the excerpt from Isaiah Bowman's *The New World: Problems in Political Geography* suggests, this vision challenges the civilizational narrative conveyed by the fault line and instead suggests shared interests forged through political and territorial unity. Bowman had been a vociferous advocate of the establishment of a pan-Slav state in southeast Europe, and had played a key part in establishing the state through his role in the negotiations of the Treaty of Versailles in 1919 (see Ó Tuathail, 1994; Smith, 2003). Significantly, Bowman's analysis rests both on the similarities between peoples within Yugoslavia and on the necessity for unity in the face of 'formidable' neighbouring states. Barrier geopolitics, then, develops the fault line by opening the possibility of forms of solidarity between peoples within Yugoslavia on the basis of its position between antagonistic political forces.

There are a number of implications to this conceptualization of space. First, this discourse centres on the role of state territoriality as a pacifying force capable of mediating diverse interests through a confederated model. Rather than basing contemporary politics on ancient cleavages, this approach foregrounds the state in Weber's (1958) terms as the holder of a monopoly of legitimate violence and hence political differences may be negotiated with civility rather than through force. Second, the belief in the protective and stabilizing power of state borders illustrates the conservative nature of the inter-state system, where rights of self-determination are often denied on the basis of preserving existing state territoriality (Agnew, 1994; Elden, 2009). This conservative impulse is reflected in the initial desire in 1991 and 1992 among external observers to attempt to preserve Yugoslav state territoriality, and later to base future state boundaries on the existing boundaries of the six Yugoslav republics. The wish to protect Yugoslav territoriality was based on security concerns that reached beyond the Balkan Peninsula, reflecting the insecurities of state elites in sanctioning action that may set precedents for other territorial disputes. Third, the barrier analogy inserts the territory of Yugoslavia into the wider historical drama of the break-up of the Ottoman and Austro-Hungarian empires and its subsequent positioning within Cold War geographies. A barrier cannot exist on its own (it would not be barring anything) but is relationally conceived as an obstacle between territories. Within this optic, changing global hierarchies and relationships necessitate a shift in conceptualization of Yugoslavia. Fourth, and perhaps most importantly, the barrier narrative shares with the fault line script (and other bold geopolitical pronouncements) a faith in stark geographical storylines through which to understand the making and unmaking of states. This reflects a classical geopolitical perspective that writes global space in such a way as to enable certain political and territorial interventions. Of course, such policy prescriptions may operate in opposition to those enabled through the 'fault line' analogy, and it is in the tension between these understandings of Bosnian political space that plural understandings of the state emerge.

Like many classical geopolitical storylines, the barrier narrative masks more than it reveals. In order to understand the social and cultural capital of this term there is a need to examine how the concept of the 'barrier' emerged over the nineteenth century and then found particular currency during the latter part of the twentieth century. This was not a unitary or teleological process, but rather a reaction by a disparate set of actors located within and beyond the borders of Yugoslavia to changing global political dynamics. In order to understand this process we need to engage with the historical and geographical details that lie behind the establishment of the Yugoslav state. The birth of the first Yugoslav state (initially named the Kingdom of Serbs, Croats and Slovenes) at the Corfu Declaration in 1917 was the result of anti-imperial struggles by activists loyal to Serb and Croat causes, and as such these proponents saw state sovereignty as a moderating force that might unite disparate political interests. In this sense Yugoslavia was, from its earliest incarnations, the site of a collective anti-imperial politics, if not a shared political identity. As we will see, the 'barrier' image gained particular currency during the early decades of the Cold War, where the stability and prosperity of Yugoslavia became a strategic interest to a range of political elites in both the West and East. Yugoslav leaders later translated this positioning 'between' Cold War power blocs into political capital through the establishment of the Non-Aligned Movement. In what follows, this narrative of unity, antagonism and statehood is explored through six historical moments.

3.2.1   Illyrianism and the First Yugoslavia

The emergence of the discourse of a unitary Yugoslav state can be traced to the fall of the Austro-Hungarian Empire following the First World War. The latter part of the nineteenth century had seen growing support for the unification of the Southern Slavs, or Yugoslavism, through visions of national liberation, modernization and the development of characteristics deemed unique to South Slav culture on the Balkan Peninsula (Prpa-Jovanović, 2000: 44). This movement had emerged among Croatian intellectuals in the 1840s as 'Illyrianism' (after the earliest tribe known to have inhabited the Balkan Peninsula) and called for the ethnic and political unity of the Southern Slavs. There were, however, varied motivations for Yugoslavism among the two largest Slav groups, the Croats and the Serbs. For Croat intellectuals, Yugoslavism formed a central strand of a broad goal of stimulating a Croatian cultural renaissance in order to overthrow Austro-Hungarian rule (Cohen, 1995: 2). In Serbia, the Prime Minster Ilija Garašanin set out his motivations for the support of Yugoslavism in the 1844 publication *Načertanije*, or Memorandum. This document, which only became public in 1906, called for the unification of all areas of the Balkans with large Serbian populations, amounting to a blueprint for an expanded 'Greater Serbia' (MacKenzie, 1996: 216). For Serbia's politicians, then, the idea of South Slav unity before the First World War was a useful concept only to the extent that it might facilitate and hasten the achievement of Serbia's distinct national and territorial goals (Cohen, 1995: 6–7).

With growing support for Yugoslavism, albeit with potentially conflicting motivations, new alliances in the early twentieth century began to alter the geopolitical power bases within the region. First, in the Balkan Wars of 1912–13 the Balkan League (comprising Serbia, Montenegro, Greece and Bulgaria), with the support of Russia, drove the Ottoman forces out of Europe and then went on to divide the territory of Macedonia. Following this success against the Ottomans, self-confidence among the Serbs ran high. In the words of Rebecca West (2001 [1941]: 355), they 'saw themselves cutting loose from the decaying corpse of an empire and uniting with a young and triumphant democratic state'. These victories seemed to erode Croatian and Slovenian commitment to the Yugoslav idea, as the growing strength of Serbia in the Balkans was viewed with suspicion. While still calling for the unification of the Slavs, it was suggested that this could take place through the Hapsburg monarchy, thus avoiding direct union with the Serbian state. These calls represent an interesting mobilization of imperial, as opposed to state, sovereignty as a practice through which competing nationalisms could be accommodated.

The question of Slav unity was to leave an indelible mark on both European and global geopolitics through the assassination of the Hapsburg Archduke Franz Ferdinand in Sarajevo on 28 June 1914. The perpetrator, Gavrilo Princep, was a member of *Mlada Bosna* (Young Bosnia), 'a loose association of youthful and nationalistic Bosnian Serbs' (Mackenzie: 1996, 247). Austria-Hungary did not wait long to respond and, amid accusations of sponsoring the assassination, declared war on Serbia on 28 July 1914. As Germany supported the Austro-Hungarian cause, and with Great Britain, Russia and France rallying to the Serb side, this theatre of conflict marked the beginning of the First World War. The violence of this conflict, which pitched the Southern Slavs against each other, saw the loss of an estimated 1 900 000 lives in the region. However, the war also resulted in the collapse of Austria-Hungary, leaving the right political conditions for the unification of the Southern Slavs. Such an outcome had become a reality through the Yugoslav Committee, an exiled collection of Croat, Serb and Slovene politicians, whose work culminated in the Corfu Declaration of 1917 calling for a democratic constitutional monarchy uniting Serbs, Croats and Slovenes (Prpa-Jovanović, 2000: 49). This declaration was the precursor to the formal recognition of the Kingdom of Serbs, Croats and Slovenes on 1 December 1918, later renamed Yugoslavia.[2]

Despite considerable internal tensions (see Jović, 2009), the First Yugoslavia remained a unified state up until the outbreak of the Second World War. Between 1939 and 1940 Germany placed considerable political pressure (with the threat of violence) on the Balkan Peninsula, primarily with a view to Romania's valuable oil supplies (Glenny, 1999: 458–459).

Yugoslavia maintained a neutral position in these early years of the conflict and, as such, became the focus of considerable lobbying and enticements from both the Allied and Axis powers. Early 1941 saw a change of tactic, as the German invasion of Romania and Hungary increased insecurity within Yugoslavia. As an attempt to prevent an attack, Yugoslavia signed up to the Tripartite Pact with Germany. However, this act provoked a coup d'état on 27 March 1941, led by the head of the Yugoslav Air Force, General Dušan Simović. This popular uprising met little opposition from forces loyal to King Aleksandar, the head of the Yugoslav state, as 'they saw no point in resisting' (Glenny, 1999: 474). Rather bizarrely, while this coup had been based on opposition to the Tripartite Pact, once General Dušan Simović seized control he decided, perhaps pragmatically considering the army ranged against them, to honour the pact. However, this was too late for the Führer who had become 'irritated by the intricacies of Balkan politics' and ordered the bombing of Belgrade on 6 April 1941 (Glenny, 1999: 476).

Hitler and Mussolini partitioned Yugoslavia less than two weeks later, dividing its territory between Germany and Italy (Prpa-Jovanović, 2000: 57). Following the creation of the Independent State of Croatia (*Nevavisna Drzava Hrvatska* or NDH) across an area roughly equatable with present-day Croatia, BiH and Slovenia, pro-Axis Ustaše ('Insurrectionists') emerged out of the extremist wing of Croatian nationalism and attacked Serb Četniks ('Irregulars'), led by Dragoljub 'Draža' Mihailović and loyal to the king (who, along with the rest of the Serbian royal family, sought exile in London). However, in Belgrade a rival resistance force began to emerge from the Communist Party, a negligible and persecuted political presence during the 1920s and 1930s. This group, led by the half-Croat, half-Slovene Josip Broz under the *nom de guerre* Tito,[3] urged partisan warfare (gaining them the name 'the Partisans') and sought to defend the ideal of Southern Slav unity. This ideology was one factor in their successful resistance: it constituted a moderate alternative to the ultra-nationalist causes of the Croat Ustaše and the Serb Četniks. In addition to their broad support base, the Partisan movement also gained the backing of the Allied forces, traditionally supporters of Draža Mihailović's Četniks. In *Eastern Approaches* (1949) the then British Secret Service agent Fitzroy MacLean describes this changing allegiance:

> Information reaching the British Government from a variety of sources had caused them to doubt whether the resistance of General Mihailović and his Četniks to the enemy was all that it was made out to be. There were indications that at least as much was being done by armed bands bearing the name Partisans and led by a shadowy figure known as Tito. Hitherto such support as we had been able to give had gone exclusively to Mihailović. Now doubts as to the wisdom of this policy were beginning to creep in. (MacLean, 1949: 279)

Figure 2   Map of Yugoslavia in 1990
Source: Author, adapted from Glenny, 1999: xx

The Allied support for the explicitly Communist Partisans involved 'swallowing ideological scruples' (Zimmerman, 1996: 6), particularly as the Serb political elite had been staunch supporters of Great Britain in the First World War. Nevertheless, 'they weren't killing the most Germans' (MacLean, 1949: 281) and following the capitulation of Italy in September 1943 supply lines were established across the Adriatic to assist Tito's Partisan forces.

On 29 November 1943, the wartime Partisan parliament convened in the Bosnian town of Jajce, liberated from the German occupiers, and voted for a new Yugoslavia that would, in its final form, be a federation of six republics, with two autonomous provinces, Kosovo and Vojvodina, joined to Serbia (see Figure 2). Under Allied pressure, the exiled royal government backed the Jajce Declaration and, in September 1944, broadcast an appeal to 'all Yugoslav patriots to rally to Tito' (Singleton, 1976: 96). Mihailović made frantic efforts during 1944 to wrest the political initiative from Tito, particularly in the form of the proposals emerging from the Ba Congress of January 1944. However, these attempts were destined to be unsuccessful in the face of growing international recognition and acceptance of Tito's Partisans as a government-in-waiting. Indeed, Soviet troops entered Yugoslavia from Romania and Bulgaria and assisted the Partisans in liberating Belgrade on 20 October 1944.

Tito was left with a dual manoeuvre to instil a sense of a coherent Yugoslav state. First, he sought to neutralize domestic military enemies. During 1945 Tito struck at Ustaše and Četnik forces with savage brutality. He used these attacks to justify the rapid growth and unchecked authority of the Partisans' wartime security service, the *Odeljenje za zastitu narodna* (Department for the People's Defence or OZNa), with an explicit remit to 'strike terror into the hearts of those who did not like this sort of Yugoslavia' (Lampe, 1996: 223). The focus of this purge of opponents, initially fixed on pro-Axis war criminals, shifted to include suppression of what it viewed as 'reactionary nationalistic tendencies', particularly in Croatia and Kosovo (Tepavać, 2000: 65). It is estimated that between 1945 and 1947 as many as 100 000 people across Yugoslavia were executed, with many more interned for their wartime activities or political views (Lampe, 1996: 223). Second, and in tandem, Tito sought to establish a sense of solidarity among the population of the Yugoslav state through the popular slogan *bratstvo i jedinstvo* ('brotherhood and unity'). Tito's approach emphasized a policy of equality between the nations of Yugoslavia and equalizing the blame for the atrocities carried out by the Ustaše, Četnik and Nazi forces during the Second World War. The commemoration of the conflict was based more on the heroics of the Partisans against a 'fascist enemy' than on seeking to present the multiple lines of difference that fractured the population over the five years of war (Judah, 2000).

3.2.2   Tito's Yugoslavia

To confirm, rather than to threaten, Tito's grip on power, elections were held in 1945 with the professed intention of granting the Yugoslav population the opportunity to vote on their country's future. In reality, and unsurprisingly considering the activities of the OZNa in crushing nationalist opposition through internment and summary execution, the ballot box choices in this postwar election amounted to a vote for Tito's Communist Party or, if you were keen to express opposition, ticking a blank box. Other political parties had either been disenfranchised (often for alleged wartime collaboration with the Četniks or Ustaše) or had excluded themselves from the election out of protest (Lampe, 1996: 226). Following the election, Communist representatives dominated the Constituent Assembly which first convened in Belgrade on 29 November 1945. The delegates voted unanimously to abolish the monarchy, ending the regency of the exiled King Petar II, in whose name Tito had ruled since March of that year (Lampe, 1996: 229). The new Federal People's Republic of Yugoslavia (SNRJ) now took its place. Tito drew up a constitution mirroring closely Stalin's Soviet model of 1936, while making specific clauses to help foster his vision for Yugoslavian unity. Malcolm (1994) points out how attempts to devolve power to the six republics were contradicted within the constitution itself:

> It [the constitution] contained the usual mixture of fine-sounding declarations and logical black holes, proclaiming, for example, that each constituent republic was 'sovereign', but also eliminating the right to secede by declaring that the peoples of Yugoslavia had decided to live together forever. (Malcolm, 1994: 194)

In reality the constitution of 1946 established a Yugoslavian state that was, initially at least, highly centralized. The ruling principle of the constitution was that of the dictatorship of the proletariat, exercised by the ruling party in the name of the workers. The controlling hand of the Socialist Party was evident at all levels of political, economic and cultural life, though the party itself was not mentioned in the constitution (Singleton, 1976: 106). This socialist rhetoric was matched by a radical reorganization of the postwar Yugoslavian economy with the nationalization of industry and, albeit at a slower pace, collectivization of agriculture. In the early years of the new Yugoslav state, Stalinism was enforced with great brutality (Glenny, 1999: 551). However, the enthusiasm for Soviet methods of 'iron rule' masked a growing schism between Tito and Stalin. There are many different opinions as to the roots of the split between these two leaders. Some, such as Allcock (2000), primarily view the conflict as an economic issue emerging out of Yugoslav dissatisfaction with the terms of trade between their country and the Soviet Union. Others see the dispute as a reflection of Stalin's general disapproval of the independence and self-confidence of Yugoslavia, and in particular his anger at criticisms by members of the Yugoslav government concerning the conduct of the Red Army in Belgrade (in particular their treatment of the Yugoslav population) during the final days of the Second World War (Tepavać, 2000: 66). Whatever the underlying cause, the rift between the two countries led to a terse exchange of letters over 1948 that resulted in the ejection of Yugoslavia from the Cominform (Communist Information Bureau).[4] Lydall (1984) summarizes the accusations levelled at the Yugoslav government in these letters:

> The Yugoslav Party leaders were accused of deviation from Marxism-Leninism, of being hostile to the Soviet Union and the Soviet Party, of borrowing from 'the arsenal of counter-revolutionary Trotskyism', of becoming nationalists, of ignoring the class struggle in agriculture, of disguising the role of the Party, of refusing to accept criticism, and of 'boundless ambition, arrogance and conceit'. (Lydall, 1984: 62)

It is clear from the reaction of the Yugoslavian leadership that the split with Stalin was unexpected, and the almost instant economic impact left the country's five-year plan, instituted to deliver rapid industrialization, in disarray. The Soviet bloc attempted to teach the 'arrogant and conceited' leadership a lesson by enforcing an economic blockade which, by 1950, had halted all trade between Yugoslavia and Eastern Europe (Lydall, 1984: 63). In the face of economic crisis, Tito was faced with a dilemma. On the domestic front he was keen to increase the pace of agricultural collectivization, 'demonstrating that it was not Yugoslavia but the Soviet Union and its allies that had strayed from the path of Stalinist orthodoxy' (Glenny, 1999: 547). However, in order to survive, Yugoslavia had to move closer to the West to attract foreign aid and foster trading links. An attempt to meet these apparently contradictory domestic and international objectives was forged through a reappraisal of centralized Soviet practice. This process involved denouncing the Soviet socialist system as a 'bureaucratic deformation' of Marxist principles due to the vast centralized power of the state over the economy (Lydall, 1984: 62). In order to circumscribe the centralized role of the state, a system of 'workers' self-management' was proposed in 1949, suggesting a move away from centralization and towards workers' councils and self-managed enterprises.

These developments provoked a distinct post-Stalin Yugoslavian foreign policy. In light of its greater independence, coupled with the signs of a more participatory form of governance (albeit with a lingering lacuna for human rights abuses), Western powers saw the benefit of closer ties to this European socialist anomaly. In recognition of this new geopolitical position, Tito helped establish the Non-Aligned Movement, eventually launched in 1954–55 with member-states from across Asia, Africa and Latin America. This movement was 'non-aligned' in its broad neutrality between the capitalist and Soviet powers. This position of independence, while open to the entire world, proved vital in preventing a Soviet attack on Yugoslavia while also encouraging overseas aid, trade and investment (Tepavać, 2000: 71). Non-alignment and greater proximity to the West meant that during the 1950s overseas aid underpinned an ambitious scheme of Yugoslavian industrialization (in favour of other sectors such as agriculture), while also subsidizing consumption, 'especially that of the state' (Allcock, 2000: 75). While these aspects were to lead to economic problems in the future, such as balance-of-payment deficits and inflationary pressure, the Yugoslav economy grew faster from 1953 to 1961 than most others in the world, including those of the Soviet bloc (Lampe, 1996: 272). These developments were replicated across Yugoslavia, and by 1961 industrial production was increasing by 13% per annum, leading commentators to talk in economic terms of a 'great leap forward' (Dyker, 1990: 28).

However, this macro industrial optimism masked two underlying economic and social phenomena that, together, are thought to have contributed to the fall of Yugoslavia: increasing rural-to-urban migration and the growing economic disparity between the six republics of Yugoslavia. In terms of the migration from rural areas, from 1948 to 1971 the ratio of rural to urban population throughout the country changed from 70:30 to 40:60 (Singleton, 1976: 222–223). By the 1980s this phenomenon was widely being described as a 'rural exodus' (McFarlane, 1988). Much of this migration can be attributed to the ambitious industrialization programme, which led to a decline in the proportion of the economically active population engaged in agriculture, from 67% at the end of the Second World War to 38% by the census of 1971 (Allcock, 2000: 89). The migration of large sections of the population from rural to urban areas was more than an economic phenomenon. Slavenka Drakulić (1996) describes the rural-to-urban resettlement in Eastern Europe following the Second World War as a 'giant leap from feudalism to communism' (Drakulić, 1996: 36). Glenny (1999: 628) suggests that Yugoslav society was particularly polarized as a consequence of its exposure to Western Europe which 'accentuated the cultural divisions between rural and urban life as nowhere else in the Communist Balkans'.

The second phenomenon that tempered the optimism of growing Yugoslavian industrial prosperity was the widening economic inequality between the six republics. The southern republics of Macedonia, Serbia (in particular the autonomous province of Kosovo), Montenegro and BiH never had the economic success of Slovenia and Croatia, and under Tito's stewardship this inequality was growing. As an example, in 1952 gross domestic product (GDP) per capita in BiH was 95.5% of the Yugoslav average, but by 1960 it had declined to 76% (Allcock, 2000: 83). While Croatia and Slovenia enjoyed proximity to the lucrative Italian and Austrian markets, sustained by the legacy of historic links to the Habsburg Empire, areas such as Kosovo were poorly industrialized and still reliant on agricultural production. Tito's regime had resisted investing in agriculture (beyond the bureaucratic requirements of collectivization), preferring instead to appease vocal impoverished communities with the construction of 'political factories', that is, enterprises located more for their political expedience than economic efficiency. Such considerations are evident in the example cited by Lampe (1996: 276) of the Montenegrin plant designed to manufacture refrigerators located on a mountain top and only accessible by an unpaved road that was impassable for the better part of the year. The reason for its precarious positioning was the unwavering support by the local population for the Partisan cause during the First World War and, perhaps equally important, Montenegro was a significantly overrepresented republic in the Communist Party leadership.

3.2.3 Decentralization

As the republics urbanized and became increasingly insular over the 1960s, calls increased for a loosening of the centralized federal government and increasing 'republicanization' (devolving power to the six republics) of the Yugoslav state. There was an emergent resentment among the Croatian economic elite, benefiting as they were from tourism along the Dalmatian coast, who felt that Croatia was 'milked of the fruits of its economic success in order either to support economically dubious projects in the underdeveloped regions or to subsidise government profligacy in Belgrade' (Lampe, 1996: 90). These concerns solidified into wider calls for greater autonomy for the Croatian people and, echoing the claims made in the last days of the Kingdom of Serbs, Croats and Slovenes in the 1930s, accusations of subjugation of Croats under the political control of a Serb-dominated Yugoslavia. Tito characteristically crushed this 1971–72 movement, known as the 'Croatian Spring', and its principal participants (mainly Croat scholars and academics) were imprisoned. Tito faced similar problems in Serbia where, in the wake of alleged oppression of Serbs by the majority Albanian (Kosovar) population in Kosovo, nationalist stirrings were causing unrest and rejection of Yugoslav principles of 'brotherhood and unity'. Again, this uprising was forcibly curtailed, though the sentiments of greater republican autonomy and calls for a democratization of the Yugoslav system were more difficult to silence.

Tito's solution lay in further amendments to the constitution and, finally, an entirely new Yugoslav constitution in 1974. This document, running to 406 articles (at the time the world's longest constitution), established a regimented, and highly confusing, pyramid hierarchy with Tito at its head. At the local level, six groupings (workers in the social sector; peasants and farm workers; liberal professionals; state officials and soldiers; local communities; and socio-political organizations) were eligible to send delegates to the next level above, the Assembly of the Commune (*opištine*) and the Assembly of the Republic. These bodies would then send delegates to the next level – the Federal Assembly. The notion of representation through delegates, as opposed to electoral competition and universal suffrage, was championed by Tito who suggested that 'a determined break has been made with all the so-called representative democracy which suits the bourgeois class' (Tito, 1974: 46). In contrast, Lydall (1984) suggests that the delegate system appeared to be designed to stifle potential opposition:

> Under the system of delegations and the pyramid structure to higher bodies, coupled with the complete ban on organised opposition, there is no possibility of a candidate with an independent viewpoint and popular support entering a commune assembly, let alone the Federal Assembly. (Lydall, 1984: 126)

In addition to the delegate system, the constitution devolved significant power to the republics, giving each a central bank and separate police, educational and judicial systems. It also gave these functions to Kosovo and Vojvodina which became 'autonomous provinces' within Serbia. Perhaps as a reaction to the recent nationalistic tendencies, these measures circumscribed Serb power within Yugoslavia, and the constitution was perceived to 'cut

Serbia down to size' (Silber and Little, 1996: 34). Significantly for BiH, the constitution granted Bosnian Muslims the status of a constituent nationality [*narodna*] of Yugoslavia (together with the Slovenes, Croats and Serbs). While this status did not translate to any concrete political representation within the Bosnian republic (in which they had a plurality), it did introduce specific measures for cultural and linguistic protection within the Yugoslav federation (a right that was conspicuously not extended to the Kosovar Albanians). While such constitutional reform would appear to indicate devolution of power from the centre, Tepavać (2000: 72) suggests that in reality centralist power was 'simply transferred to the level of the republic'. In addition, the rotating system of representation at the federal level, established to 'prevent any single republic or politician accumulating too much power' (Glenny, 1999: 623), proved confusing and was poorly understood by the Yugoslav population (Lampe, 1996: 306).

As the 1970s progressed, the worsening economic position of Yugoslavia was to overshadow constitutional debates. The economic decline of the federation was masked in the short term by the import of capital (and the associated build-up of debt) and the export of labour, particularly in the form of *gastarbeiter* (guest workers) to Austria and Germany. In the wake of global oil price rises in 1973–74 and 1978–79, interest repayments on overseas loans increased, and, as unemployment rose in Western Europe, *gastarbeiter* began to return to Yugoslavia seeking work. The political changes necessary to alleviate the worsening economic position at the federal level looked increasingly difficult considering the frailty of Tito's leadership and the complex and over-bureaucratized constitution. Somogyi (1993: 46) suggests that by the late 1970s the federation was being run by 'informal, non-elected, non-institutionalized, and uncontrollable oligarchies'. Following Tito's death in 1980 the infighting between the republics continued, and in his place the rotating presidential system (as established by the 1974 constitution) further entrenched political stagnancy. Yugoslavia's precarious economic position began to impact upon the inflation rate over the early 1980s, and by the end of 1984 it had reached an annual rate of between 60% and 100%, requiring some lower-income Yugoslav families to spend 70% of their monthly incomes on food (Ramet, 1985: 6).

In this political and economic climate, the principles of 'brotherhood and unity' were again under threat. In 1986, the Serbian Academy of Science and the Arts leaked a Memorandum to a popular newspaper, a document that claimed that the Serbs were in an unjust position in Yugoslavia and were being exploited by Slovene and Croatian compatriots. In a phrase that would be repeated often over the following decade, the Memorandum asserted that Serbia had 'won the war but was losing the peace' (Silber and Little, 1996: 31). It went on to denounce the 'creeping genocide' inflicted on the Serbs in the 'cradle of Serbian culture', Kosovo (Zolo, 2002: 19). This call-to-arms unsettled the Yugoslav leadership, who almost unanimously denounced the document; one notable exception was Slobodan

Milošević, then a young Socialist Party chief and friend of the Serbian President Ivan Stambolić. As we have seen in the discussion of 'fault line' geopolitics, Milošević was to see first-hand the power of nationalistic discourse in winning the support of Serbs, first in Kosovo and then across Yugoslavia. Following a televised parliamentary showdown with Stambolić in 1989, Milošević gained the Serbian leadership. He also managed to engineer the dissolution of the autonomous provinces of Vojvodina and Kosovo. This meant that, with Montenegro, he held four of the eight votes in the Yugoslav parliament. The other four republics (Slovenia, Croatia, BiH and Macedonia) lacked any unity of opposition to Milošević's creation of what Misha Glenny (1999: 628) describes as 'Serbo-slavia'.

Just as 'fault line' geopolitics supported certain interventions in Yugoslavia and BiH, the scripting of Yugoslavia as a barrier 'between' political entities in southeast Europe shaped political agendas in a number of ways. The first is that the barrier lent coherence and homogeneity to the Yugoslav state, when the centrifugal tendencies of competing republics, uneven economic development and intricate constitutional manoeuvrings suggest a more fragmented reality. The second implication was that the 'barrier' concept was mobilized in the early 1990s to support a range of different political agendas. For example, the actions of Milošević are suggestive of a hollowing out of barrier geopolitics where the state is retained as the primary territorialization of political life, but repurposed to stand as a container of an ethnically defined population. Here the barrier and the fault line are concurrently deployed, where difference is essential and transhistorical but may be addressed through an alternative state territorialization. As David Campbell (1998a) has argued, this project forges a connection between identity and territory, where political difference is negotiated through spatial exclusion rather than agonism. But this is only one of the implications of barrier geopolitics, since the response to the fragmentation of Yugoslavia saw reticence first by many international observers to recognize successor states, and then secondly to enshrine the borders of the six republics (and later Kosovo) as acceptable state boundaries.

3.3   The Balkan Vortex

The Balkans were the original Third World, long before the Western media coined the term. In this mountainous peninsula bordering the Middle East, newspaper correspondents filed the first twentieth-century accounts of mud-streaked refugee marches and produced the first books of gonzo journalism and travel writing, in an age when Asia and Africa were still a bit too far afield. [...] Twentieth-century history came from the Balkans. Here men [sic] have been isolated by poverty and ethnic rivalry, dooming them to hate.

Kaplan (2005: ii)

When considering a unifying narrative for his travels across southeast Europe, Robert D. Kaplan follows others in settling on an imagined geography of 'the Balkans'. With a faded sepia cover of images of the mosque-lined streets of the Kujundžiluk neighbourhood of Mostar projected over an ancient map of Constantinople, Kaplan's book draws on travels across southeast Europe to explore the historical antecedents to conflict and social fragmentation. As the excerpt suggests, Kaplan has no hesitation in framing violence as a product of a conflict-ridden space, trapped in a history of inter-ethnic violence. For Kaplan, the mere mention of the 'Balkans' suggests antagonisms that do not conflate with Western forms of political contestation. Such a narrative is drawing on an established literary device that has conveyed an imagined geography of the Balkans as a space of deviance, conflict and enmity positioned at the margins of Europe. The identification of this literary trope has led scholars to apply the critical tools of Said's (1978) *Orientalism* to representations of the Balkans, founding a new object of critique: *Balkanism*. Over the past two decades Balkanism has emerged as a distinct form of discursive critique, isolating the power relations masked in representations of Balkan identities and locations. For Maria Todorova (1997), critiques of Balkanism draw attention to the mechanisms and registers through which the Balkans have served as a 'repository of negative characteristics against which a positive and self-congrat-ulatory image of the "European" and the "West" has been constructed' (Todorova, 1997: 188). Within this discourse 'Europe' stands for modernist ideals of rationality, morality and consensual politics while the 'Balkans' are cast as a place of barbarism, irrationality and 'ancient hatreds'.

The 'Balkan vortex' is the final geopolitical framework considered in this chapter. As with both the 'fault line' and 'barrier' imaginaries, this spatial narrative of the Balkans naturalized a set of cultural and political assumptions about the nature of Yugoslav society and hence the most appropriate forms of intervention in its fragmentation. The role of Balkanist tropes within media, political and literary fields has been widely debated within critical scholarship examining international interventions in Yugoslavia (Dodds, 2005; Helms, 2008). But one aspect of this imagined geography has received less attention. Balkanist imagineries do not simply imply an inert space within southeast Europe that is a site of primordial deviance. There is also a suggestion within such accounts that this disposition has the capacity to draw in others from outside. Just as fault line geopolitics has drawn on the field of tectonic geomorphology, Balkanist accounts have looked to fluvial processes to convey this attraction. Consequently, Balkanist accounts often describe this area of Europe as a 'vortex' or 'whirlpool' (see Ó Tuathail, 1996). The Balkans do not form a passive backdrop to artistic representations; the location is itself drawn into the enactment of deviance.

There are many examples of such centripetal representative strategies. Bram Stoker's *Dracula* (2003 [1847]) opens with Jonathan Harker's account of his journey from London to Castle Dracula in Transylvania. These pages are laden with Balkanist imagery of an unknown and unmapped destination, far from the civility of life in London. But the account is also one of the most explicit examples of the sense that this is a region that has an undefinable attraction, that it is actually drawing those in from outside. For example, as Harker describes the topography of the Translyvanian region:

> I read that every known superstition in the world is gathered into the horse-shoe of the Carpathians, as if it were the centre of some sort of imaginative whirlpool; if so my stay may be very interesting. (Stoker, 2003 [1847]: 6)

For Vesna Goldsworthy, whose book *Inventing Ruritania* (1998) contains a comprehensive discussion of the use of Balkanist imagery through Western art and literature, such representations serve as a means of subverting 'a variety of taboos and satisfying hidden desires' (p. 126). One of the interesting points here is the simultaneous deployment of the unknown and of seductive attraction. Agatha Christie's fictional work is replete with examples of such a combination of spatial uncertainty and attractive energy. For instance, in *The Secret of Chimneys* (1925) Christie used the fictional Balkan land of 'Herzoslovakia' as a homeland for the villain Boris Anchoukoff, a country where the national 'hobby' is 'assassinating kings and having revolutions' (Christie, 1925: 105). The use of a fictional name, a fusion of Herzegovina and Slovakia, seems to suggest that the Balkans are 'so hopelessly and intrinsically confused and impenetrable that there is scarcely any point in trying to distinguish between them' (Fleming, 2003: 1). The use of aspects of Slav nomenclature is enough to convey to the reading public the moral corruption of certain characters or places. But Christie also looks to use the Balkans as a specific backdrop to the enactment of crime. For example, *Murder on the Orient Express* (1934) mirrors Stoker (2003 [1847]) in focusing on a train journey into the Balkans and a parallel descent into criminal deviance. The title's murder takes place when the train comes to a halt between Belgrade and Zagreb, literally in a space between Istanbul's oriental bazaars and the apparent safety of Italy's piazzas. But the mobility is crucial here: the characters move from the imagined 'Orient' into uncertainty, immobility (the train is stuck in a snowdrift) and threat.

Balkanist styles of representation are not restricted to Western art and literature. They have been used within literature from 'Balkan' countries to illuminate the prejudices that shape Western representation. For example, in *The Days of the Consuls*, originally published in 1941, Nobel Laureate Ivo Andrić (2000 [1941]: 24) describes the reaction of a young French consul on arrival at the Bosnian town of Travnik at the beginning of the nineteenth century:

> The land was wild, the people impossible. What could be expected of women and children, creatures whom God had not endowed with reason, in a country where even the men were violent and uncouth? Nothing these people did or said had any significance, nor could it affect the affairs of serious, cultivated men.

This statement is an ironic reflection on French attitudes towards BiH; since Andrić was born in Travnik, this is a self-description read through an agent of French colonial rule. But this is only part of the significance of Andrić's account. The book describes seven years of the consul's life spent in Travnik, detailing the rather mundane bureaucratic obligations he was obliged to fulfil. It is a story of mobility and immobility, variously illustrated by his initial arrival, his period of service and his final departure. The Balkans, within this optic, is again seen as incorporating a form of malevolent attraction.

3.3.1   Conflict in BiH

The representation of the Balkans as a vortex played a prominent role in both domestic and international interpretations of the fragmentation of Yugoslavia between 1991 and 2008 (Močnik, 2005; Ó Tuathail, 2002). Balkanist representations of deviance provided a flexible set of imaginaries which could be used to support a range of policy positions, by both leaders of emergent successor states in Yugoslavia and Western intervening agencies. Balkanist language allowed policy makers to perform two manoeuvres. First, to present the violence as an inevitability, stemming from the specific history and geography of the region. But such inevitability is not restricted to environmental context, since the language of ethnic hatred orientates attention on the biological constitution of the population. Ethnic hatred, inscribed through the blood circulating within individual bodies, simmers beneath the surface only to 'bubble up' (more fluvial terminology) at intermittent points in history (Bose, 2002: 11). Second, the presentation of the Balkans as a vortex suggests the threat of immobility posed by intervening in the violence. As Ó Tuathail (1996) discusses, in the initial phases of the BiH conflict US politicians were wary of being sucked into Vietnam-style engagement. But not only does a vortex convey a sense of being drawn into a long-term military commitment, it also suggests the difficulty of identifying the forces through which the violence is sustained. Without a credible 'exit strategy', international policy makers were wary of sanctioning intervention (Rose, 1998).

These two manoeuvres were evident in international interventions in the fragmentation of Yugoslavia. As Croatia and Slovenia declared independence in 1991, the initial response of political elites across Europe and the United States was to explore the legal basis of new sovereignty arrangements. This centred on legal precedent, with the European Economic Community establishing the Arbitration Commission of the Peace Conference on the former Yugoslavia (or Badinter Commission) in August 1991 in an attempt to set some legal framework for the establishment of new states. The recourse to international law illustrates an attachment to 'barrier' geopolitics, where territorial status quo played a central part in the arbitration process of the Badinter Commission (see Radan, 2000). But from the outset the attempt to rely on international legal convention to prevent conflict was flawed. The unilateral declarations of independence by Slovenia and Croatia in 1991 solicited a violent response by the *Jugoslovenska narodna armija* (Yugoslav People's Army or JNA) as Milošević struggled to hold the Yugoslav state together. Milošević's act can be (and was) presented both as a virtuous attempt to retain a multi-ethnic Yugoslav state and as ethnically motivated violence seeking to keep all Serbs in a single state (see Judah, 2000).

We can see similar multifaceted explanations for the violence that broke out in BiH in 1992. In line with the criteria set by the Badinter Commission, BiH held an independence referendum on 29 February and 1 March 1992. The result was an overwhelming vote in favour of independence, though the turnout was only 63% as Bosnian Serbs had boycotted the vote, branding it 'illegal' (Zimmerman, 1996: 188). One month earlier, Serbian politicians, led by Radovan Karadžić's SDS, had declared the creation of the Republika Srpska, proclaiming it part of Yugoslavia (Silber and Little, 1996: 218). Following the declaration of independence their strategy militarized, as Serb forces, including the JNA, erected barriers around Serb villages in BiH and demarcated Serb areas in Sarajevo. One month later the United States and the then European Community recognized BiH's independence, citing the majority vote at the referendum. This act played into the hands of Radovan Karadžić who claimed that, as in the past, the outside powers were bent against Serbian sovereignty (Udovički and Stikovać, 2000: 179). Through discourses of victimhood and threats to Serb national security, violence spread as paramilitaries and the JNA combined to expel non-Serb residents from the newly created RS.

The violence in BiH stimulated a range of responses from international commentators. In the UK, politicians often described the conflict as 'bewildering' or a 'moral sickness', as the events in BiH were attributed to 'ancient ethnic hatreds' or 'primordial evil' (Major, 1999; Owen, 1998; Robinson, 2004). For example, the then UK Prime Minister John Major used the 1994 Lord Mayor's Speech in London to celebrate the end of the Cold War but identify new security threats:

> The lights have come on again all across Europe. As they have done so, they have illuminated some dark corners of the Cold War: ethnic tensions, seen at their worst in the country that used to be called Yugoslavia. (Major, 1994: np)

The juxtaposition of a 'light' Europe with a 'dark' Yugoslavia recalls the Balkanist tendency to present geographical difference in a simple binary between the virtuous self and the deviant other. But Major was not alone in using such language. Famously, US President Bill Clinton is said to have read Kaplan's *Balkan Ghosts*, and this experience shaped US policy towards intervention in the conflict in BiH (Ó Tuathail, 1996: 212). Certainly, Clinton's inaugural presidential speech on 21 January 1993 made reference to the threat posed to the post-Cold War world of 'ancient hatreds' (see Clinton, 1993), and subsequently over the early months of his first presidential term the United States moved away from the 'lift and strike' policy which advocated lifting the Yugoslavia-wide arms embargo and striking Serb military positions in BiH.

In the case of the spatial articulations of both Major and Clinton we must not grant too much coherence to their policy positions, nor should we view them as a definitive break from 'fault line' or 'barrier' geopolitics. Major's geopolitical reasoning views Balkanist tensions as a product of the collapse of the barrier positioning of Yugoslavia between Cold War adversaries. The presentation of Balkanist discourse occurred alongside a demographic *realpolitik* underpinning initial support for Serb causes. This is captured by Lord Carrington's remark that the Yugoslav population were 'all impossible people … all as bad as each other, and there are just *more* Serbs' (cited in Simms, 2001: 17). Caution also needs to be used when causally connecting Kaplan's writing with Clinton's iteration of Balkanist discourse. As Hansen (2006: 122) points out, we cannot be certain of the extent to which reading *Balkan Ghosts* exerted an influence over Clinton, since the narrative of Balkan historical enmity in Kaplan's book mirrors security discourses circulating within US government circles at this time. Nor should we overstate the coherence of Kaplan's argument; his writing meanders away from Balkanist imagery, and includes gestures towards 'fault line' geopolitics and the nostalgia for a multi-ethnic past.

As discussed in the following section, the representation of the violence in BiH in Balkanist terms by domestic and international politicians had a profound effect on the nature and institutions of international intervention. The history of Western literature and art provided a repository of spatial descriptors, resources that were used by political elites to improvise responses to the violence. Significantly, the geopolitical framework of the vortex or quagmire provided the justification for avoiding large-scale Western troop deployment or strengthening the mandate of UN troops already on the ground in BiH.

3.4   The General Framework Agreement for Peace

In the years since the signing of the GFAP, members of the PIC have undertaken repeated attempts to broker its renegotiation. The reason for the dissatisfaction with the original agreement, and its attendant BiH Constitution, is that it accommodated contradictory ideals within a single state structure. We see in the GFAP attempts to variously endorse the spatial imaginaries of the fault line, the barrier and the vortex. Indeed, it is through a desire to hold in tension these three geopolitical narratives that the paradox of the GFAP emerges. The agreement itself marked the end of four years of diplomatic efforts to conclude the violence, largely enacted through a series of European conferences and proposed peace agreements. Across these practices two frameworks were used by domestic and international politicians to explain the violence. The first centred on national security, the second humanitarianism. Considering each in turn illuminates the origins of the Dayton paradox.

The initial political and military confrontations in BiH in 1992 were framed by nationalist leaders (both within BiH and in other former Yugoslav republics) in terms of competing understandings of national security, structured around mutual fear of minority status in new state territorialities. Political groups representing Bosnian Muslims (or Bosniaks) and Croats feared Serb political dominance in what remained of the Yugoslav state (following the secession of Croatia and Slovenia in 1991). In contrast, Serb-nationalist political parties, in particular the *Srpska demokratska stranka* (SDS) led by Radovan Karadžić, feared minority status in an independent BiH. These fears fuelled antagonistic political rhetoric which transformed into violence following the Bosnian government's declaration of independence in April 1992 (see Silber and Little, 1996). Over the following months Serb military and paramilitary forces, supported by the well-armed Yugoslav People's Army (JNA), attempted to create an ethnically homogeneous territory within BiH, called the Republika Srpska (or RS) (Kadrić, 1998). This action involved the expulsion or execution of non-Serb populations, the besieging of key cities such as Sarajevo and Goražde, and the holding of prisoner populations in a series of camps in Prijedor, Brčko and Bijeljina, among others (Dahlman and Ó Tuathail, 2005a). Similar atrocities, if on a smaller scale, were committed by groups claiming to represent Bosniak or Croat national interests, in particular in the southern Bosnian towns of Konjic, Mostar and Čelebići.

The second discursive framework was humanitarianism. Western political elites portrayed the violence as a humanitarian disaster (Ó Tuathail 2002), focusing attention on the plight of Bosnian citizenry, in particular in the besieged streets of Sarajevo. This biopolitical label drew attention to the outcome of violence and the failure of the state to perform its function of protecting its citizenry, but it simultaneously erased a conception of victim and perpetrator. Presenting the conflict in these terms allowed for an institutional response that focused on the distribution of aid and the promotion of NGOs. The humanitarian label was sustained through explanations by politicians in the United States and the UK that the violence was a consequence of 'ancient ethnic hatreds' or 'primordial evil', labels that present conflict as biologically predetermined and an essential part of a Balkan temperament rather than as a political manoeuvre. Silber and Little (1996) lament how foreign diplomats 'behaved as though the war had no underlying structural causes at all [...]. They behaved as though all they had to do was to per-suade the belligerents of the folly of war' (Silber and Little, 1996: 159). Crucially, such Balkanist interpretations of the conflict led to an assumption that the only means of resolution of the violence was the partition of territory along ethno-national lines. Drawing on Jacques Derrida (1994), David Campbell (1998a) describes this alignment between territory and identity as 'ontopological', as national identities are fused with the particular territories (Derrida, 1994 in Campbell, 1998a: 80).

Reflecting the widespread adoption of discourses of national security and humanitarianism, the primary tool of conflict resolution was the map. From the initial attempts at the Lisbon Conference of February 1992, through the Vance Owen Plan of 1993, to the Dayton GFAP in 1995, international diplomats based their interventions on the possibility of finding the 'correct' division of Bosnian territory that would prove acceptable to all sides in the conflict. This approach represented a paradox: negotiators were willing to accept the internal division of BiH into ethno-national regions but were, at the same time, committed to retaining the boundaries of the Bosnian Republic as the official borders of a multi-ethnic state (Cox, 2003: 256).

Following a ceasefire arrangement in 1994, Bosniaks and Croats had grudgingly agreed to create a Federation, a move that simplified the efforts to establish a universally acceptable division of Bosnian territory (see Morrison, 1996). The pace of negotiation was further increased over 1995 as NATO air strikes against Serbian military positions led to territorial gains for Bosniak and Croat forces, thereby encouraging Serb representatives back to the negotiating table (Holbrooke, 1999). The GFAP, negotiated over late 1995 by US Ambassador Richard Holbrooke, involved the partition of BiH into two sub-state 'Entities' divided by the Inter-Entity Boundary Line (IEBL): the Muslim-Croat Federation[5] (with 51% of Bosnian territory) and the RS (with 49% of the territory). The civilian implementation of the GFAP was to be overseen by the internationally led Office of the High Representative (OHR), with military implementation provided by a 60 000 strong NATO implementation force (I-For). Bosnian government competencies were largely devolved to the Entity level (and from there below to local *opštine*); the central state institutions were left weak and suffered from complex power-sharing arrangements between the three main ethnic groups[6] (Bojičić-Dzelilović, 2003). Despite the centrifugal nature of the GFAP constitution, Annex VII of the agreement called for the return of all displaced persons (DPs) and refugees to their pre-war homes (see Dahlman and Ó Tuathail, 2005b). This demonstrates the paradox at the heart of the GFAP: nationalist cartographies were sanctioned through the creation of ethno-national spaces, while coexistence and multi-ethnicity are simultaneously endorsed through the promise of refugee return.

While the Dayton negotiations had involved a painstaking process of assigning land to either of the Entities, in the carefully balanced ratio of 51:49, the fate of the northern municipality of Brčko remained a source of strong disagreement between the negotiating parties. The demography of Brčko had altered dramatically over the course of the conflict, as military forces loyal to Serb causes had expelled the predominantly Bosniak population of the town. Within the geography of the Dayton state, Brčko municipality had accrued a political significance as the link point between the two halves of the RS (ICG, 2003: 2). Bosniak and Croat delegates argued that giving Brčko to the RS would reward the ethnic cleansing while also denying the Federation access to the River Sava. Serb negotiators countered that the RS would not be viable in two parts, arguing that a continuous and defendable Serb territory was a condition of their signature. Consequently, resolving the fate of the town became the 'toughest of all issues at Dayton' (Holbrooke, 1999: 296). Following intense negotiation the stalemate was finally broken when Slobodan Milošević agreed to put Brčko under international arbitration, with a decision reached within one year (Silber and Little, 1996: 376). As will be discussed in the following chapter, this proved an ambitious timetable as the arbitration process ultimately took over four years and was shaped by shifting local, regional and international political contexts.

3.5   Conclusion

This chapter has explored the history of BiH through three geopolitical narratives: the fault line, the barrier and the vortex. Each of these histories attempts to explain the political features of the present through recourse to the geographies of the past. Rather than providing a linear history of BiH, I have sought to present an account that traces the origins of many of the popular narratives concerning BiH that inflect present-day political interventions. This is most apparent in the characteristics of the GFAP, where the retention of the borders of the Bosnian state betrays an enduring belief in state territoriality to mediate social difference, while also illustrating the fears on behalf of international negotiators of setting a precedent of formally partitioning the Bosnian state. The reliance on state territoriality would suggest an enduring attachment to 'barrier' geopolitics, where a unified Bosnian state may protect its citizenry from incursions from more powerful neighbours. But simultaneously, the geography of the GFAP illustrates the underlying attachment to 'fault line' interpretations. Negotiators considered the partition of Bosnian territory into ethno-nationally derived spaces (the RS and Federation, and then within the Federation into cantons) as the only mechanism to stem the violence. This compromise between state territoriality and partition provides an initial example of state improvisation, a politics of the possible worked out across two weeks of proximity talks on an airbase in the US state of Ohio. The performance continued with the signing of the GFAP in the Palace of Versailles on 14 December 1995, a gesture that was presumably designed to recall the 1919 Treaty of Versailles and the subsequent reorganization of the European state system. This initial example of using the symbolic capital of the Western states to convey the legitimacy of the Bosnian state through recourse to historical precedent and stately pageantry demonstrates the projection of the imagined timeless quality of state power, performed through ephemeral practices.

The following chapter develops this analysis of statehood through an examination of attempts to resolve the anomalous position of Brčko following the signing of the GFAP. This example provides an opportunity to explore the contradictions and ambiguities inherent in state improvisations. The creation of Brčko was reliant upon a range of transnational actors, policies and networks, though its ultimate aim was to produce a form of rule that appeared unitary, sovereign and shaped by the will of the local population. While this recalls the inherent ironies of interventions to foster democratization, it also allows us to explore the forms of tactics and performances that are utilized to attempt to convey the state in precarious times.

Notes

1 For historical accounts of the establishment and fragmentation of Yugoslavia see Glenny (1999) or Udovički and Ridgeway (2000); for accounts of the conflict see Silber and Little (1996).

2 This state is often referred to as 'the first Yugoslavia' to act as a distinction from the socialist 'second Yugoslavia' formed in 1945.

3 MacLean (1949) recalls how Tito gained this name on account of continually telling people what to do: '"you," he said to them, "will do this; and you, that"; in Serbo-Croat, "Ti, to; ti, to." He said this so often that his friends began to call him Tito' (MacLean, 1949: 311).

4 Stalin's postwar coordinating body of nine European Communist parties.

5 Despite the unity of the Federation on paper, to a certain extent its division into ten cantons acknowledged the *de facto* existence of a Croatian third entity of Herceg-Bosna, stretching west from the Herzegovinian town of Mostar to the border of Croatia (Klemenčić and Schofield, 1996).

6 One exception to this was the Constitutional Court of Bosnia and Herzegovina, which retained a significant role in mediating in disputes between the two Entities (see Bose, 2002: 61–68).

Chapter Four

Performing Brčko District

Seeing has been identified by a range of scholars as central to the production and reception of the idea of the state (Corbridge *et al.*, 2005; Scott, 1998). While I was conducting fieldwork in BiH, public visual displays by multiple international agencies formed a backdrop to daily routines. These were often spectacular: the armoured personnel carrier from the US Army Base Camp McGovern outside Brčko cruising through the centre of town on regular patrols; or the establishment of the Office of the High Representative (OHR) offices in a former school surrounded by a newly constructed eight-foot fence. But they were more often mundane: the billboard poster celebrating UN intervention, the United States Agency for International Development (USAID) sticker on a refurbished youth centre, or the United Nations Development Programme (UNDP) logo on the side of their Local Action Programme vehicles. Nor was the use of such symbols restricted to international organizations. Local NGOs often gauged the success or failure of a meeting or seminar on the basis of the quality and visibility of their institution's name and logo. Through these varied displays there seemed a perpetual desire among different organizations to brand the landscape, to demonstrate the impact of their work by visual means. But I would argue that the use of logos within reconstruction projects exceeds simply informing passing pedestrians or participants of donor largesse. Like all brands, they are drawing on the power of image or phrase to convey a set of qualities and emotions. It appears that repetition of these state-building brands constitutes part of a practice to challenge the temporary nature of project work and consultancy, to suggest instead uniformity across a variety of projects and interventions across time and space.

The Improvised State: Sovereignty, Performance and Agency in Dayton Bosnia, First Edition. Alex Jeffrey.
© 2013 John Wiley & Sons, Ltd. Published 2013 by John Wiley & Sons, Ltd.

As we will see, improvisations of the state extend beyond such branding. However, these examples demonstrate an important aspect of the character of such improvisations. Performances of the state may be dynamic and transient, but they are often attempts to fix authority to stable and lasting concepts and locations. They constitute ephemeral moments which seek to claim permanence. This book has advanced a theory of state improvisation as a combination of performance and resourcefulness, therefore each performance seeks to draw on resources that are suggestive of durability and intransigence. As Fiona McConnell's (2009) work illustrates, such suggestions of spatial and historic uniformity are central components of a state's claim to legitimacy. In the rapidly transforming context of Bosnian political life such improvisations necessitate a double manoeuvre: to produce new laws, institutions, processes and norms while simultaneously invoking a sense of stability. This tension is only exacerbated by attempts to claim that new state practices not only are uniform across a specified sovereign territory, but reach back through time and recover established historical practices, identities or positions. It is through these claims to historical precedent that we see the significance of the multiple interpretations of Yugoslav history covered in the previous chapter. As policy makers, NGOs, politicians and activists have struggled to narrate a coherent national narrative for BiH, they have variously drawn on geopolitical images of the fault line, the barrier and the vortex. As we saw, these storylines are contradictory, so their deployment illustrates the divergent understandings of present and future state arrangements in BiH.

Such plural and conflicting performances of the state will be explored through an examination of state-building strategies in Brčko District. As set out in the previous chapter, a decision on the fate of the Brčko municipality was omitted from the GFAP on the grounds that disagreement over this territory risked destabilizing the entire peace agreement. This led to a protracted process of arbitration, where the fate of the municipality was negotiated between representatives from the United States (acting on behalf of the Peace Implementation Council), the Republika Srpska (RS) and the Muslim-Croat Federation (Federation). After numerous delays, the outcome was the creation of Brčko District in 1999, a unique territorial status in BiH where the territory of the two Entities notionally overlapped. This resolution prompted an increase in international funding, particularly from the United States, and the establishment of a new post within the OHR of District Supervisor. Through these intense processes of international funding and new governmental strategies we can observe how international and local officials sought to communicate the idea of Brčko District as a local state, within BiH. There are aspects of this analytical framework that render Brčko *sui generis* in terms of the wider experience of other Bosnian municipalities. Many local or international commentators viewed the anomalous position of Brčko with suspicion, or even felt that it was undeserving of 'District status'. As one OHR Official in Sarajevo remarked, 'I can think of five municipalities in BiH that are more in need of District supervision.'[1]

It is clear that the Brčko District 'model' offers an alternative political resolution to others attempted at the GFAP (Dahlman and Ó Tuathail, 2006; Mukić, 2007). But as set out in the introduction, the concern of this book is not whether or not Brčko reflects wider Bosnian experience, since scaling up observations made at municipal level is always going to raise concerns regarding comparability and generalization. Rather, the Brčko experience provides an intensive and highly localized attempt to establish new sovereign arrangements in BiH. While the empirical focus may be Brčko District, the story that unfolds illustrates a common refrain in state-building contexts: the plural performances of state authority that coexist across the same territory. The enactment of Brčko District required changing individual understandings of the nature of statehood, to make people 'think' in terms of a new idea of the state.

This chapter is divided into five sections. The first provides an overview of the history of Brčko, detailing the forms of violence, exclusion and reconstruction that have taken place in the years since the GFAP. This set of narratives illustrates the significance of Brčko, as a contested territory and site of intervention. The second, third and fourth sections each examine one of the central narratives of state improvisation: stability, security and neutrality. In terms of 'stability', I draw on evidence gathered from political elites in Brčko, in particular within the OHR, that sought to present the new District arrangements as a form of 'stable' multi-ethnic government that reflected the ethnic make-up of the pre-war Brčko population. A set of improvisations were performed by the OHR to convey such a multi-ethnic democratic polity: from imposing new education policies to recruiting a new District Assembly and Council. As we will see, these practices, designed to demonstrate the stability of the new District structure, were prone to crisis and reversal as alternative imaginings of the state came to the fore. The third section explores how ideas of 'security' were mobilized in Brčko District to justify economic reform. Attending to the war-ravaged economy of the District was the highest priority for the OHR, and this process deployed a particular understanding of security centring on the protection of foreign capital. This process of neoliberal reform has constituted a fundamental part of state improvisation, as a range of local and international agencies have been drafted into reforming government budget processes and delivering services. The fourth section examines how the creation of the District involved the reorganization of the symbolic landscapes of the former municipality to imply neutrality. Street names, changed during the conflict, were renamed following the establishment of the District to reflect a multi-ethnic future, while new commemorations such as 'Brčko District Day' sought to cultivate new forms of solidarity (see Hobsbawm and

Ranger, 1983). Across all three of these sections we see the need to explore state improvisation empirically, since there is considerable dynamism between official discourses of the state and their implementation in everyday life. While these state performances are suggestive of stability and uniformity, terms in their improvisation we see the significant 'grey space' between official and unofficial, legitimacy and illegitimacy, legality and illegality (see Yiftachel, 2009).

4.1   Brčko District and Arbitration

The Brčko municipality covers 493 square kilometres of northeast BiH, extending from the slopes of the mountain of Majevica in the south to Brčko town beside the River Sava in the north (see Figure 3). The municipality largely consists of rich agricultural land and the town benefits from a large port on the Sava providing a trade connection to the Danube in Belgrade and further to the Black Sea basin. The town was bitterly contested during the Bosnian war, as Serb paramilitaries and the *Jugoslovenska narodna armija* (Yugoslav People's Army or JNA) occupied the town for the entire duration of the conflict. This aggression was part of the wider attempt by Serb forces led by the *Srpska demokratska stranka* (Serb Democratic Party or SDS) to establish security for their minority population within BiH through the creation of an ethnically homogeneous Republika Srpska (RS). As Dahlman and Ó Tuathail (2005b) point out, this philosophy of assuring national security through homogeneous territory contradicted the 'actually existing fabric of everyday life and the ordinary domicile security of a functioning multiethnic Bosnia' (Dahlman and Ó Tuathail, 2005b: 577, drawing on Bringa, 1995). Prior to the conflict Brčko exemplified this heterogeneity: the 1991 census states that the Brčko municipality had a population of 87 672, of whom 44.07% were Bosniak, 20.69% were Serb, 25.39% were Croat, 6.54% declared themselves Yugoslav, and 3.3% others. The town had a population of 41 406, of whom 55.53% were Bosniak, 19.93% were Serb, 6.99% were Croat, 12.58% declared themselves Yugoslav, and 4.96% others (Kadrić, 1998: 21). As this data suggests, Brčko was consistent with other areas of Yugoslavia in having a strong representation of 'Yugoslavs' in the urban area, denoting individuals that felt their strongest affiliation not to a national group but to a multi-ethnic identity. The data also illustrates that the town, though multi-ethnic, had a majority Bosniak population, while the rural areas consisted of a tapestry of Serb, Bosniak and Croat villages.

The demographic configuration of Brčko municipality changed dramatically during the conflict. The position of Brčko as a link point between the two halves of the RS (see map inset), as well as providing a supply route to the Serb Krajina region of Croatia, led to the town having particular strategic significance for Serb forces. The importance of Brčko to

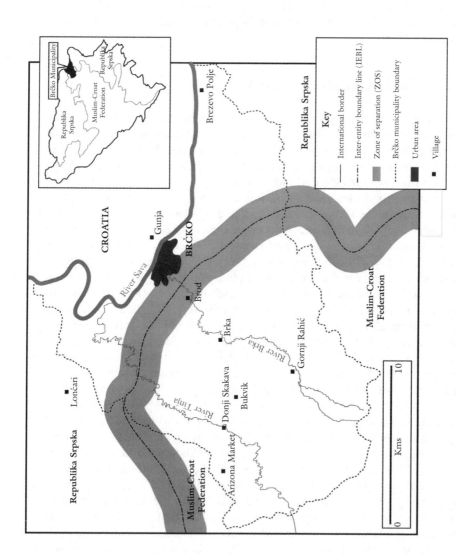

Figure 3 Brčko Municipality and BiH after the GFAP
Source: Author

the logistic success of a 'greater Serbia' translated to a series of atrocities against the Bosniak and Croat inhabitants of the town over the period of Serbian occupation. Following the Serb invasion in April 1992, Bosniak and Croat women and children were forcibly expelled from Brčko, with Bosniaks seeking refuge in the villages of Brka and Gornji Rahić and Croats in Dornji Skakava. From these locations Brčko residents re-established their *mjesne zajednice*[2] (local community associations or MZs) 'in exile' to continue to function as focal points for the community. As a consequence of these expulsions, Brčko municipality split into three sub-municipalities: 'Brčko Grad' (Brčko town) housing an exclusively Serb population and, south of the frontline, the Croat 'Ravne-Brčko' and the Bosniak 'Brčko-Rahić'. Many Bosniak and Croat men and boys were unable to leave Brčko town and were instead held in informal collection areas such as the central police station, the bus company '*Laser*', the hospital, the Džedid Mosque and the Brčko *luka* (port) (Human Rights Watch, 1992; Kadrić, 1998). A number of testimonies exist suggesting that war crimes took place at each of these sites,[3] though the most brutal massacres were carried out at the *luka* camp, leading to two men (Goran Jelisić and Ranko Češić) being indicted by the International Criminal Tribunal for the former Yugoslavia (ICTY) for crimes against humanity.

In addition to targeting the population, the Serb forces in Brčko also attacked the urban fabric of the town as symbols and spaces of cultural heterogeneity were destroyed. Described variously as 'domicide' (Porteous and Smith, 2001) or 'urbicide' (Coward, 2002, 2004), this process of targeting urban spaces has been seen as part of the ethno-nationalist programme to 'eradicate difference in order to create and naturalise the idea of separate, antagonistic sovereign territorial identities' (Coward, 2004: 266). The most explicit examples of urbicide in Brčko were the destruction of the large hotel in the centre of town, the removal of numerous Yugoslav monuments, the razing of all four mosques within the town limits and the renaming of the street network to reflect the Serb occupation of the town. For example, the main traffic route through Brčko was renamed '*Bulevar Đenerala Draže Mihajlovća*' (after the Second World War Četnik leader Draža Mihajlović), the central shopping street '*Srspskih Oslobodilaca Brčkog*' ('The Serb Liberation of Brčko') and the road out of town towards the western RS and the Krajina '*Krajiški Put*' ('Krajina Way'). Thus many historical connections between Bosniaks or Yugoslavs and the town were either obscured or removed as a reconfigured landscape celebrated newly invented connections with Serbian history and mythology.

In the context of the Bosnian war the experiences of Brčko are not unique. From a pre-war Bosnian population of 4.4 million, 1.5 million people became refugees across a total of twenty-five countries and around one million were internally displaced over the period of the conflict (Dahlman and Ó Tuathail, 2005b: 14; UNHCR, 1997a: 30). In addition, many urban areas across BiH were subjected to similar processes of urbicide, targeting heterogeneity and the cultural symbols of victim populations (see Bose, 2002; Coward, 2002, 2004; Robinson *et al.*, 2001). Rather, the importance of Brčko emerges in the unique approach taken by international mediators resolving the conflict in the municipality both during and after the GFAP negotiations. Drawing on fieldwork undertaken in three fieldwork periods between 2002 and 2005, this chapter will examine the strategies through which international agencies have attempted to re-create a cohesive local state across the fractured and divided post-conflict Brčko municipality. Through such a qualitative focus the multiple practices of municipal-level statecraft emerge, as local and international actors have collaborated to communicate the 'idea' of a unified and multi-ethnic Brčko. These aspects of international intervention in Brčko can only be understood in the context of the wider international response to the Bosnian conflict.

4.1.1   The Brčko Arbitration

The International Arbitral Tribunal for Brčko was led by former legal adviser to the US State Department Roberts Owen, with Professor Čazim Sadiković (representing the Federation) and Dr Vitomir Popović (representing the RS). Though Annex II of the GFAP stated that the area under arbitration was illustrated on an attached map, no map had in fact been drawn, and consequently the first task for the Arbitral Tribunal was to decide on its territorial remit. The Tribunal decided to arbitrate over the entire pre-war Brčko *opština*, fractured as it had been since 1992 into three sub-municipalities. Similarly to the various negotiations during the Bosnian conflict, the arbitration process between 1996 and 1999 was shaped by events 'on the ground' in Brčko, as well as by the broader geopolitical context of international intervention in BiH. The pronouncements of the Tribunal were directly influenced by the progress of GFAP implementation in Brčko and to this end the negotiations seemed to focus on two main indicators: the return of refugees and the holding of democratic elections.

At the end of the war the United Nations High Commission for Refugees (UNHCR) estimated that there were 52 333 displaced persons in Brčko *opština*: 35 073 in Federation parts (Ravne-Brčko and Brčko-Rahić) and 17 261 in the RS part (Brčko Grad) (UNHCR, 1997b). Over the initial postwar period the displacement of the population continued in Brčko, as RS authorities drew on the vacant housing stock to house displaced Serbs from Sarajevo, Glamoć, Jajce, Sanski Most, Bihać and the Croatian Krajina. As with other areas of BiH, it was the two kilometres of territory either side of the IEBL, which following the GFAP was known as the Zone of Separation (ZOS), which proved most contentious in Brčko (see Dahlman and Ó Tuathail, 2005b; Toal and Dahlman, 2011). The ZOS covered a significant area of the Brčko suburbs, including neighbourhoods such as Omebegovaća, Dizdaruša, Gajevi and Brodusa which had previously been home to majority Bosniak and Croat populations. There was concern among the RS authorities in Brčko that the demographic alterations brought about by returns to the ZOS could weaken their claim to the municipality within the arbitration process: they were 'reluctant to alter the facts on the ground'.[4] This concern provoked a two-way reconstruction effort over the summer of 1996, with both Federation and RS authorities attempting to fill the vacant housing in the Brčko ZOS with members of 'their' ethnic group (Griffiths, 1998). In addition to this race to fill vacant houses, other tactics of 'micro-level humiliation and contempt' (Dahlman and ÓTuathail, 2005b: 21) were mobilized against returnees, such as the obligation to hold an RS identity card with the Serbian military symbol of the twin-headed eagle.

As the returns process threatened the fragile peace within the Brčko area, American I-For, located in the new Camp McGovern army base in the ZOS, declared a two-week moratorium on construction in the ZOS over July 1996. In this period, the OHR (in conjunction with the UNHCR, IPTF[5] and I-For) established the International Housing Commission (IHC) to screen return applications to ensure the rights of the claimants to the properties. While there was a clear need to oversee the return of families to their pre-war homes, the complex bureaucracy of IHC, coupled with the explicit lack of a guarantee of security to returning families, undermined the international efforts. Consequently, incidents of damage to property or individuals were common over this period of the IHC, as documented by the International Crisis Group (ICG):

> On the 11th November 1996, 9 reconstructed houses were dynamited in Brod and Omerbegovaća. Between the 28th of February and 11th March 1997, 11 newly pre-fabricated houses were destroyed in Gajevi. In all, 200 houses owned by displaced Bosniaks were either blown up or burnt down before the IHC programme was halted. (ICG, 1997: 2)

The Arbitral Tribunal took particular interest in these events as the activities of RS officials were deemed to have failed to comply with their obligations as laid out at the GFAP. The focus of the Tribunal on these events led Gojko Klicković, then prime minister of RS, to pull out of the arbitration proceedings in December 1996 (ICG, 1998: 3). However, when the formal arbitration hearing began in Rome in January 1997 the RS changed its mind, perhaps concerned that the entire Brčko municipality would be handed over to the Federation. The outcome of the Rome negotiations, issued on 14 February 1997, was an interim decision (a final resolution was deferred for another year), with the delay blamed on ongoing failures in terms of 'freedom of movement and the return of former residents to their Brčko homes' (Arbitral Tribunal, Article I in OHR, 2001).

The strategy selected for ensuring GFAP implementation in the future was a radical scaling-up of international intervention in Brčko. The cornerstone of this approach was the formation of a new OHR office in Brčko (OHR-North), headed by a Deputy High Representative for Brčko, otherwise known as the 'Brčko Supervisor'.[6] In order to break the stalemate that surrounded the political, economic and legal realms in Brčko, this post was granted a wide-ranging set of powers at the February 1997 Rome Declaration 'to supervise Dayton implementation [and] strengthen local democratic institutions' (OHR, 2000: 258). In March 1997, in a further supplementary award published at a PIC meeting in Vienna, High Representative Carl Bildt appointed US diplomat Robert Farrand as Supervisor of Brčko for one year, with deputies from Russia and the UK. The first significant declaration by Ambassador Farrand was a 'Procedure for Return to Brčko', a document that established a new Returns Commission in Brčko. This new Commission had a positive impact on the rate of returns, in particular speeding up the processing of claims and, more importantly, introducing a new Brčko ID card without Serb symbols on the front cover. By January 1998, the Returns Commission had approved the return of 2461 families, the majority of which were Bosniak. Of those, 710 had actually 'returned', the criterion of which was spending at least one night at the property (ICG, 1998: 8).

Despite these successes, during 1997 a series of symbolic barriers to returning Bosniak and Croat families appeared within the Brčko townscape. On 8 September 1997, a concrete statue of Draža Mihajlović was unveiled in the centre of Brčko (fittingly on *Bulevar Đenerala Draže Mihajlovća*), though he had no connection with the town during his lifetime. Two weeks later, a twelve-foot memorial dedicated to the 'Serb Defenders of Brčko' was unveiled in an adjacent park (see Figure 4). These strategies were clear attempts to continue the 'Serbianization' of Brčko and use any means necessary to forge an historical link between the Serbian people and the townscape (a tactic that had been deployed during the GFAP negotiations[7]). In comparison to violent intimidation or the imagery of the RS ID cards, the OHR initially found these intangible symbolic aspects of the Brčko landscape difficult to legislate against. Like earlier infringements, these attempts at 'non-military ethnic cleansing' were taken into consideration by the Arbitral Tribunal.

The GFAP stated that elections would take place in BiH no later than nine months after the agreements came into force (GFAP, Annex III article III in OHR, 2000: 39). Considering the political, economic and social conditions in BiH, described above, this was an extremely ambitious timetable. A number of commentators have suggested that the reason for this compressed electoral timetable was that the BiH elections would coincide with the US presidential election and consequently allow a tangible demonstration of BiH's democratization (Donais, 2000). Despite this

Figure 4 Monument to the Serb defenders of Brčko
Source: Author's photograph

opportunity for a foreign policy triumph, the early elections suited the nationalist political parties as their control over the key domestic institutions and agencies such as police and media 'ensured that they determined the agenda and discourse of the election campaign in 1996' (Griffiths, 1998: 61). In the RS, the reporting by official Serb media was deemed so offensive and biased to the ruling SDS party that High Representative Carl Bildt accused them of broadcasting propaganda that 'even Stalin would be ashamed of' (ICG, 1996: 15 in Donais, 2000: 240; see also Human Rights Watch, 1996: 2).

In addition to the dominance of the wartime nationalist parties, the dispersed population meant that voter registration was a central problem, since voters had the option of voting either where they resided in 1991, where they resided after ethnic cleansing in mid-1996 or where they intended to reside in the future (Campbell, 1998a: 222). This principle granted opportunities for electoral fraud, as electoral registers within the RS were manipulated to include the maximum number of Serbs (Donais, 2000). Indeed, radio broadcasts in the RS suggested that those who planned to vote in their pre-war places of residence were 'directly attacking the

Serbian nation' (Campbell, 1998a: 222). This strategy was also employed within Serbia itself, where some 31 000 were 'assigned' to vote in Brčko, while an additional 20 000 were registered in the formerly Bosniak-majority town of Srebrenica (Donais, 2000: 241). Within this context, it was not surprising that the 1996 elections simply 'gave the democratic stamp of approval to the three nationalist parties that had waged the war' (Woodward, 1997: 97).

A final arbitration announcement was scheduled for March 1998 and, in light of the continued GFAP non-compliance by the RS in Brčko, there were predictions that the entire municipality would be handed to the Federation. While recognizing the moral arguments to making this shift, the Arbitral Tribunal resisted, citing the new political climate in the RS and the prospect of future returns and freedom of movement. Prior to the declaration of March 1998, Milorad Dodik had outlined in a highly influential testimony to the Arbitral Tribunal that he felt that in the future the 'IEBL will be an irrelevant issue' in Brčko (Supplementary Award March 15th 1998 in OHR, 2000: 265).

History has shown that Dodik's moderate approach, which garnered such international support, has since shifted to populist nationalism, for example through recent attempts to call a referendum on RS independence and advocate for a withdrawal from state institutions (see Chivvis, 2010; ICG, 2011). But at this stage in the Brčko negotiations Dodik's performance of moderation was a crucial intervention. The dismantling of the political relevance of the IEBL was a central aim of the GFAP, and the Arbitral Tribunal saw that a final decision on Brčko could destabilize the fledgling administration of Dodik and thus undermine his moderate stance. This calculation reflects the way in which the OHR could use the Brčko decision as conditionality towards reform in the RS; hence a swift resolution could jeopardize the implementation of the GFAP in other parts of BiH. This was also true of the Federation, where the dismal returns situation in Sarajevo was cited as a reason why a final decision in their favour was inappropriate at that time (ICG, 2003: 6). Following this award, the Bonn Peace Implementation Conference granted the same executive and legislative authority to the Brčko Supervisor that had been previously granted to the High Representative in Sarajevo. This comprised the power to sack any public official who obstructed GFAP implementation (that is, the returns process, the strengthening of democratic institutions and the revival of the economy) and the ability to pass any law that was perceived to help such implementation.

By the time of the arbitration meeting of March 1999, it was increasingly apparent that in addition to the intransigence of the local politicians, the uncertainty regarding Brčko's future was itself acting as a barrier to the implementation of the GFAP. The 'new political dawn' in the RS, much anticipated by the March 1998 Supplementary Award, was thwarted at the next electoral opportunity as the ultra-nationalist Nikola Poplašen defeated Biljana Plavšić's SNS party at the September presidential elections. This result placed Milorad Dodik in a difficult position as RS prime minister, and Poplašen set about attempting to destabilize his government. These actions led to the High Representative, then Carlos Westendorp, to sack Poplašen for abuse of power, coincidently on the same day (5 March 1999) as the Arbitral Tribunal announced the Final Award.

The Final Award unified the former Brčko *opština* in a neutral and multi-ethnic Brčko District of Bosnia and Herzegovina. At the heart of the award was the unification of the pre-war Brčko municipality, to which each Entity delegated all of its powers of governance (Final Award, paragraph 9 in OHR, 2000: 284). This decision meant that Brčko would nominally be part of both Entities, their territory uniquely overlapping, while the Bosnian state-level institutions would protect the interests of the District itself. This solution meant that the Entities would both 'gain' territory even as they 'lost' administrative authority (ICG, 2003: 7). This consolation was not enough to stop the resignation of Milorad Dodik in protest at what was perceived by many in the RS as the division of Serb territory in BiH. However, the resignation had little impact, as Dodik remained the *de facto* prime minister: he was the only individual with the necessary votes to form a government and the only possible interlocutor between the RS and the OHR (ICG, 2003: 8). Serb discontent was exacerbated by the NATO air strikes in the Federal Republic of Yugoslavia (FRY) in late March 1999 at the culmination of the Kosovo crisis.

In the context of Serb anger at the perceived injustice of the arbitration decision and the NATO military action, the Final Award's reliance upon the protective role of the Bosnian state over the independence of Brčko was optimistic. In reality, with the central state institutions left so weak following the GFAP, the Final Award is underwritten by a commitment for intensified international supervision to defend the interests of the District from incursions from either Entity until the state can take over this role. The District Supervisor during the 2002–03 fieldwork, US Ambassador Henry Clarke, saw Entity encroachments as one of the key future threats to Brčko: he felt that a strong District was a necessity, 'otherwise the Entities will eat Brčko alive when the Supervisor is gone. It sounds a little dramatic but if you knew how much time I spent worrying about RS encroachments on Brčko in particular then you would know why I am inclined to use strong words.'[8]

The role of the Supervisor, and that of the Brčko OHR more broadly, has been to implement the Final Award (indeed, in September 2002 the OHR Brčko office, formerly 'OHR-North', changed its name to 'The Office of the Final Award'). When interviewed, the Supervisor was clear about his duty, while attending to criticisms of his comprehensive powers (see Chandler, 1999):

Office of the
High Representative

Aleksandra Karadjordjevica 2, 76100 Brcko
Tel: 381 76 205 666 Inmarsat Fax: 871 382 040 838
Local Fax : 381 76 205 560

SUPERVISORY ORDER
ON THE ESTABLISHMENT OF THE BRCKO DISTRICT
OF BOSNIA AND HERZEGOVINA

MARCH 8, 2000

In accordance with Paragraph 9, of the Brcko Final Arbitral

Award dated March 5, 1999, and in consultation with the High

Representative, I hereby declare the Statute of the Brcko District of

Bosnia and Herzegovina to be in force.

This Order shall have immediate effect.

Robert W. Farrand
Deputy High Representative
Supervisor of Brcko

Figure 5   Supervisory order creating Brčko District
Source: OHR, 2000

I try not to appear dictatorial but everyone knows my agenda: it is the Final Award, you can go and read it. It says basically change everything, and my predecessors and I think that you cannot change everything at its present level but you have to improve it as you go along. (Interview with Ambassador Clarke, the Brčko Supervisor, 24 March 2003)

On 8 March 2000 the Brčko Supervisor released a Supervisory Order, no more than a few sentences in length, declaring the creation of Brčko District (see Figure 5). The implementation of the Final Award has since been a process of creating Brčko as a single administrative unit. In contrast to the partition sanctioned at the GFAP, this process of unification has involved establishing shared multi-ethnic practices of government over the pre-war *opština*. This was simultaneously, then, a performance of the past (unity of the Yugoslav-era *opština*) and an invocation of a shared future (the model of Brčko District).

4.2 Stability: Getting the Job Done

In February 2003 in a *ćevabdžinica* [grill restaurant] in Brčko I interviewed Beth Stevens, who was then a member of the Planning and Development Collaboration International (PADCO) team but had previously held senior positions in the OHR in Brčko. From 2002 to 2005 a group from PADCO, an NGO with headquarters in Washington, DC, carried out a programme with the District Government funded by USAID (see PADCO, 1988). This project involved establishing a District Management Team (DMT) and had three goals:

> to help the municipal administrations improve their operational efficiency, increase the quality of municipal services to citizens and businesses, and dramatically raise levels of transparency and openness in local government. (PADCO, 2005: 1)

Much of this work was targeted at reforming the budget process in Brčko and making government decisions more 'transparent' for the wider population. The programme was staffed by a group of four US workers and assisted by local staff. As with many other consultancy projects in Dayton BiH the staff had often worked in other international agencies and were returning to the country on a new assignment, with a different purpose, but shaped by the memories and experiences of previous roles.

In the fading light of the empty restaurant Beth articulated a range of challenges she had faced across these different roles, entangling reflections on the state of BiH or the different cultures of EU and US funding with stories of specific incidents from her working life. Towards the end of the interview Beth reflected on the main challenges facing the OHR in the early days of the establishment of Brčko District:

Actually creating Brčko District was the biggest problem we faced. Actually trying to establish District Law and implementing it. There were three laws working at the same time over the territory of Brčko District: Republika Srpska, Federation and Brčko District Law. It was the joining of three municipalities, two of which were recognized only politically by the Federation and not legally as, if they were, it was thought that it may diminish their claims on the whole area. We had to put all three back into one and lay the foundation for the District. We had to create the whole thing from scratch and from there begin to work on processes. So the biggest problem we faced was getting people to *think* in terms of Brčko District. (Beth Stevens, consultant in PADCO, former Executive Chief of Staff at OHR-North, Brčko, 20 February 2003)

Beth's account foregrounds two processes: the creation of law and, just as crucially, its recognition. This dualism is explained as a process, described here as its 'implementation'. Multiple legal codes existed (and continue to exist) across BiH, but it is the differential credence granted to these various frameworks through their implementation that defines their ability to shape behaviour. The social life of law is of central concern to Bourdieu (1986: 814), who sought to transcend the distinction between *formalism* (that law is autonomous from the social world) and *instrumentalism* (that law is simply a tool in the service of dominant groups). Instead, Bourdieu uses the concept of a 'juridical field' to suggest a distinct arena of social competition, using the notion of the 'field' to suggest a domain 'in which people with greater economic, social and cultural capital and with a habitus attuned to possibilities for gain tend to outwit poorer groups' (C. Jeffrey, 2010: 19). For Bourdieu the logic of the juridical field is determined by two factors:

> on the one hand, by the specific power relations which give it its structure and which order the competitive struggles (or, more precisely, the conflicts over competence) that occur within it; and on the other hand, by the internal logic of juridical functioning which constantly constrains the range of possible actions and, thereby, limits the realm of specifically juridical solutions. (Bourdieu, 1986: 816)

Bourdieu's notion of the juridical field provides a vocabulary through which to analyse the performances through which law is practised. Of course, by drawing into the realm of analysis forms of competitive struggle, Bourdieu is foregrounding the social context which constrains and shapes the production of legal norms. There are three implications of Bourdieu's assertion for the statement made by Beth Stevens. First, we can begin to unsettle Stevens's claims regarding temporality: rather than law acting as the mechanism through which new sovereign arrangements achieve consent, it may be that social consent is the precursor to the production of law. Second, and following on, the implementation of law must be understood in practice, through the subjects, sites and materials through which agents reproduce or resist emergent legal practices. Third, Bourdieu's conceptualization of law is suggestive of the possibility of multiple juridical fields, each in a competitive struggle to claim competence in the production of law.

Bourdieu's concept of the juridical field helps narrate the process through which the OHR attempted to 'make people think' in terms of Brčko District as a single territorial entity. One of the first tasks, as Beth Stevens suggests, was to create the District as a political entity with a single government. This process, undertaken using the increased powers of the OHR derived through the 1997 Rome Declaration, attempted to balance rhetoric of democratization within a process of international supervision. The Supervisor could point to the political implications of BiH's state-level democratization programme, which had, following the GFAP, been structured around demonstrations of electoral competition. Commentators later viewed the holding of elections only nine months after the signing of the GFAP as 'unwise and dangerous' on account of the nationalist outcome of the 1996 elections (Donais, 2000; Williams 1996). As discussed in Chapter Two, state improvisations are not solely performed for an audience within the state in question. In the case of the 1996 elections, these appeared to be directed towards domestic audiences within Europe and, in particular, North America to illustrate that intervention within BiH brought about democratic change. But, of course, in holding elections prior to the establishment of forms of democratic political exchange, the result was a ballot that granted democratic legitimacy to wartime nationalist political parties.

It is difficult to provide a succinct overview of the form of democratization advocated by the OHR through this early period of the establishment of Brčko. At the heart of the Final Award is an expectation that the OHR will create a 'multi-ethnic democratic government' within Brčko (see OHR, 2000: 277), and yet reaching this end goal did not always require democratic means. Certainly, the OHR avoided electoral competition, extending the suspension of elections that had been in place after the fraught vote in Brčko in 1997. At this stage, rather than promoting majority rule, the approach from the Supervisor appeared to centre on guaranteeing minority rights. In some senses this appears to be a corrective to the wider failure to guarantee minority rights in the emergence of successor states to Yugoslavia. There is also an interesting temporality to these processes; just as law requires the consent of the target population, the OHR's form of democratization assumed that institutions could not simply appear as centres of government but required the gradual establishment of trust among the Brčko citizenry. The suspension of elections therefore provides a form of protective autocracy. But underpinning such an assumption is a hidden teleology: that the establishment and operation of new institutions of government will build trust even if the public has not been directly consulted on their structure, purpose or constitution. We see in such reasoning an underlying 'barrier' geopolitics, that the construction of strong, cohesive state architecture may provide protection from opposing political forces (in this case the Federation and the RS).

The first task for the Supervisor was to establish a singular architecture of government over Brčko District. The Supervisor dissolved the former municipal assemblies of Brčko Grad, Ravne-Brčko and Brčko-Rahić and re-formed them into a District Government and Assembly. The new Government and Assembly members were chosen from these dissolved administrations, in addition to picking certain members from the fledgling civil society organizations in the town. One former OHR employee recounted how an individual had been 'awarded' a place on the District Assembly because of their efforts in establishing an NGO.[9] As suggested in the Final Award (and much like the interim Brčko Grad government of 1997), the composition of the new structures of governance within Brčko District followed a strict 'ethnic formula' (OHR, 2000: 285), with a Serb Mayor, a Bosniak Speaker in the Assembly and a Croat Deputy Speaker (reflecting the weightings of the 1991 Yugoslav census).

The power of the Supervisor to intervene in policy making was instrumental to the functioning of Brčko District Government, as politically sensitive decisions often ended in deadlock, with all sides acknowledging that if a decision could not be reached through the Assembly voting system then the Supervisor would impose it.[10] This was the case in the example of the integration of the District education systems, a top priority for the Supervisor, where he successfully imposed multi-ethnic schooling in September 2001 despite determined opposition from nationalist political parties[11] (see Bieber, 2005). A former Brčko OHR Official described this decision-making system as an important catalyst to the integration of the new District:

> Ninety per cent of the success of Brčko District specifically has been because there is no democracy here. That is why it has been successful. It is a purely pragmatic point of view. [...] Right from the very beginning, certainly from the beginning of the supervisory regime we did not seek to do things by consensus. It was just imposed. You know, 'democracy' came fairly low down the list of what was required in terms of getting the job done. (Former Brčko OHR Official, 2 June 2003)

As indicated by these remarks, popular participation in the reform of Brčko was deliberately limited in the period after the Final Award. Consequently, District elections were not held until October 2004, seven years after the last municipal vote.[12] The reason given for the delay in District elections was that the Supervisor was waiting for 'the emergence of political parties with "Brčko-based agendas"'.[13] Considering that three of the four main political parties in Brčko exhibited ethno-national affiliations,[14] this was an ambitious demand. However, this attitude demonstrates the desire that existed to shift political debate away from viewing Brčko District as a divisible or annexable territory, towards focusing attention on the specific needs of the local population.

The delay in calling elections provided the time to establish a single District administration; an early vote would almost certainly have advantaged nationalist political parties and perhaps led to a robust challenge to the tenets of the Final Award. This outcomes-based model of democratization has afforded significant successes in terms of integrating the fractured District and encouraging sustainable refugee and DP returns both to the town and to the rural areas. The reduction in the political significance of the IEBL in Brčko District over 2000 and 2001 was followed by a more gradual reduction in its social significance as the District infrastructure was harmonized over the three former municipalities. The process of tax harmonization and new laws of privatization have acted to increase the pace of privatization and by January 2004 sixteen enterprises had been privatized in Brčko District with the creation of 1112 jobs (Brčko Development Agency, 2002). The improved freedom of movement coupled with a strengthening in the District economy afforded a marked increase in return: between 2000 and 2004 there were 19 418 returns to Brčko District, nearly twice the BiH average as a proportion of pre-war population (UNHCR, 2005). In addition, the close supervision of Brčko District has allowed it to become a pioneer in BiH in terms of establishing multi-ethnic institutions such schools, the judiciary and the police.[15]

Despite the reforms that had been achieved in Brčko, the exercising of supervisory powers appeared to have a lasting impact on the popular perception of the OHR. On several occasions over the fieldwork period government officials referred to the power of the Supervisor, and even the appointed Mayor moved off a sensitive topic by exclaiming that he had to stop talking 'or the Supervisor will replace me!'[16] The Supervisor himself alluded to his control over individual careers by stating that he could find out any information he wished from the District Government 'as the guy would be in fear of his job'.[17] This fear seemed to have a profound influence on the conduct of the District Assembly, where debate shifted between poles of deadlock and agreement. When discussing key reforms there was little to gain for representatives of parties in opposition to the Final Award in attempting to reach consensus, and thus offering concessions to other nationalities and potentially facing removal from party lists.[18] In these instances stalemate would be reached and the Supervisor would have to step in to impose the law in question (as occurred with the integration of the District schools). In contrast, less important political decisions were regularly made with complete consensus and little debate, as individual members of the Assembly did not want to 'put their heads above the parapet'[19] and risk the wrath of the Supervisor.

In addition to these strategies of deadlock or acquiescence, a third strategy was to withdraw altogether from the new District institutions. The SDS employed this approach in October 2001 as they staged a boycott of the District Assembly as a protest over the reintegration of schools. When I asked an SDS representative about the boycott he shrugged and said, 'people don't want to be mixed, they want to be separate'.[20] From this position outside the new District institutions the SDS were able to openly criticize the 'internationally run'[21] District Government and Assembly with relative impunity. Thus, both the Supervisor and the SDS shared a concern for Brčko being run 'externally', either by Entity-level nationalist politicians or by the geopolitical concerns of the United States. There were fears, in particular on the part of a representative from the American National Democratic Institute[22] in Brčko, that the boycott could benefit the SDS electorally, though these were partially allayed when the SDS came second to the moderate *Socijaldemokratska partija Bosne i Hercegovine* (Social Democratic Party, or SDP) in the October 2004 municipal elections.

The limited debate (in terms of both content and participants) within the District Assembly led to cynicism among Brčko civil society organizations as to the possibility for political debate to deviate from an internationally preconfigured path. As one NGO worker explained, 'there does not seem to be much at stake'.[23] Given the delay in holding a District-wide election, this lack of public participation with the affairs of the Assembly was a concern for the Supervisor:

> The Assembly meetings are open and most of them are recorded more or less endlessly, well at least by Brčko Radio and sometimes by others. I mean, I don't go over there all the time because it would make them look like more of a rubber stamp than they already are [laughs]. But when I do go over there do I see business representatives and NGO representatives sitting in the audience watching what is going on? No I don't. Do I hear councillors discussing their meetings with people representing groups of citizens other than party meetings? Not much. (Interview with Ambassador Clarke, 24 April 2003)

As a consequence of the lack of an 'agonistic politics' (Amin, 2002), comprising open and critical debate and mutual awareness, political contestation and struggle seemed to move out of the new Assembly chamber to other institutional spaces in Brčko District. Key members of Brčko's NGOs and MZs saw greatest political capital in lobbying members of the OHR rather than councillors in the Assembly, seeing these international actors as gatekeepers to funding and legitimacy. In avoiding antagonism, the supervisory regime also served to distance civil society organizations from the Brčko Government and Assembly. Consequently, in establishing a coherent institutional structure across Brčko municipality, the OHR reaffirmed its own status as a key actor within the local state.

These powers to define what it is to 'be democratic' in post-conflict BiH have been the subject of criticism in the past, amid suggestions that 'democracy' has become 'a moral as opposed to a political category' and democratization now concerns 'societal values and attitudes rather than political process' (Chandler, 1999: 28). In the case of Brčko this power has been instrumental in unifying the architecture of government and providing protection to minority communities returning to both the urban and the rural parts of the District. However, to achieve this end the OHR has often been required to exercise legislative and executive powers, defending the new District institutions from incursions from either the Federation or the RS. In doing so it does not remove itself from the political realm, but rather becomes a key aspect of the state in Brčko.

4.3   Security: Constructing Legality

More than seven years after the end of the Bosnian war and despite some $5 billion in international reconstruction assistance, BiH's economy remains stagnant and dysfunctional, while the country is rapidly gaining a reputation not as an emerging market economy but as a lawless and ungovernable state dominated by organized crime and corruption.

Donais (2003: 359)

Prior to the fragmentation of Yugoslavia BiH was one of the most economically disadvantaged of the six republics. As Timothy Donais indicates, the economy was further ravaged by the conflict, reflected in a 65% fall in the per capita GDP between 1990 and 2000, from $10 725 to $4370 (UNDP, 2002: 11). This decrease in GDP was not consistent between the RS and the Federation, with the RS suffering a greater downturn in economic performance on account of larger army pension commitments, insufficient collection of taxes and, more significantly, the low levels of international aid following the conflict (see UNDP, 2009). As a consequence of this disparity, in 2002 the UNDP estimated that 24.8% of the RS population lived in 'general poverty',[24] as compared to 15.6% of the Federation population (UNDP, 2002: 51).

But macroeconomic indicators are not sufficient to articulate BiH's specific development challenges. The centrality of discourses of 'ethnicity' at the GFAP negotiations masked a more profound source of conflict in the former Yugoslavia: corruption and criminality structured towards ensuring economic gain. While Donais (2003) rightly illuminates the embedded nature of corruption within the political and economic life of BiH, this does not necessarily equate to mere 'lawlessness'. Rather than a zero-sum relationship between law and lawlessness, the experience in BiH appears to illuminate a series of different economic 'laws', some endorsed by international agencies, others tolerated and some legislated against. This plurality is captured in the work of Peter Andreas (2004, 2008) when he details the forms of complicity and symbiosis that developed between intervening agencies and criminal networks over the course of the conflict and post-conflict periods in BiH. As discussed in Chapter One, Andreas (2008) uses Goffman's (1959) dramaturgical metaphor of 'front' and 'back' stage to explain the different discourses and practices of complicity in corruption during the conflict. While on the 'front stage', policy makers from both BiH and international agencies would use public statements to vow to eradicate corrupt practices. However, in the 'back stage' of hotel lobbies, briefing rooms and coffee shops, deals were struck that contradicted the virtuous sentiments uttered in public.

The improvisations of the state in Brčko did not follow this precise theatrical geometry. It seems more akin to a blurring of front and back stage, as humour, exaggeration and cynicism were deployed to reflect on a set of commonly understood – but rarely openly acknowledged – truths regarding the nature of economic practices. So rather than lack of 'law', explaining economic deterioration and corruption requires a more plural vocabulary. As set out earlier in this chapter, 'law' can be understood as a form of symbolic capital that is capable of shaping individual behaviour and priorities. Therefore it would seem that it is a not a lack of 'law' but a lack of consensus around a specific set of legal codes that defines the experience of post-conflict BiH.

Owing to its location on the River Sava, Brčko had been a thriving port in the Yugoslav era, and also the location of a meat-canning factory, '*Bimeks*', a vegetable oil factory, '*Bimal*', and a large shoe factory, '*Izbor*'. All three industries had closed, or reduced production to nearly zero, following the end of the conflict. In terms of harmonization, the division of Brčko District between the RS and the Federation in the postwar period saw differential support to either 'side', with little aid or funding coming into the Serb town until after the announcement of the Final Award and considerable difficulties in equalizing levels of taxation and social welfare (Sommers, 2002). In addition to these harmonization difficulties, Brčko became renowned in the postwar period as an area of criminality and trafficking. The Brčko area had acted as a transit point not only between the two Entities, but also between BiH, Croatia and Serbia. The economic effect of this location is perhaps best demonstrated by the emergence of Arizona Market during the conflict. This is a 35-acre site 16 kilometres south of Brčko, located close to the frontlines that had been cleared of mines by UNPROFOR to allow trading, with the Pentagon apparently supplying $40 000 of the start-up costs (Andreas, 2004: 46).

This impoverished and criminalized economic situation in Brčko ensured that the Final Award provided a mandate for Supervisorial intervention:

It being clear that one of the main causes of tension in the Brčko area is its general economic depression and high rate of unemployment, all relevant international institutions are strongly encouraged to support the Supervisor in his [sic] efforts to revitalize the District's economy in the interests of reducing tensions in the area and promoting the cause of international peace. (Final Award, Paragraph 46 in OHR, 2000: 282)

This discursive tactic of forging a link between 'international peace' and the economic development of Brčko District handed full authority to the Supervisor to implement whichever strategies were deemed necessary to improve the economic conditions of the District. This approach has yielded some measure of success, as noted in a 2003 ICG report:

The establishment of fiscal discipline, a sensible and effective tax regime, and a business-friendly environment have resulted in significant foreign investment, a promising privatization program, and the highest average wages in the country. (International Crisis Group, 2003: i)

This healthy report, however, masks the increased intervention, control and management that have been required to realize such economic vitality. Economic improvement has been attempted in Brčko District at the cost of strengthening the authority of the Supervisor.

In the same way that democracy was defined by the supervisory regime in narrow terms as transparency, economic development has been cast in terms of security for private capital. The privatization scheme in Brčko had been rapid and comprehensive, with thirty enterprises being prepared for sale in 2000. By January 2004, sixteen enterprises had been privatized within Brčko, creating 1112 jobs (Brčko Development Agency, 2002). On account of the logistical difficulties of ameliorating the laws of the Federation and the RS, the Supervisor drew on his symbolic resources (in this case 'Supervisory Orders') to expedite the process. Through the release of a series of Supervisory Orders the privatization process revitalized the authority of the District Supervisor. In addition, by granting primacy to privatization as the preferred discourse of economic reform, the improvised state was also strengthened through the need for 'foreign expertise' to assist the District Government. As discussed above, this is most explicitly demonstrated by the formation in 1999 of the DMT, run until 2005 by Washington-based Development Alternatives Inc. and funded by USAID. This group assisted in the privatization process through techniques such as the preparation of 'concept papers' which were 'adopted by the international supervisory authority' (Feit and Morfit, 2002: 4).

The economic reform in Brčko demonstrates the symbolic capital of the Supervisor exercised through the ability to confer legality on certain practices (such as privatization) and illegality on others. This division between legitimate and illegitimate economic activity did not seem an easy one to make in Brčko District. The repeated use of 'black' and 'grey' economies in Brčko terminology reflects the existence of a socially constructed sliding scale between different understandings of economic practices. This seems to support Carolyn Nordstrom's (2003) assertion that 'clear distinctions between legal and illegal, state and non-state, or local and international are often impossible to make' (Nordstrom, 2003: 332). This hybridity was demonstrated in an exchange with the District Mayor (DM) during a fieldwork interview:

> DM: What we need in Brčko is a casino, perhaps in the old hotel Revena. If we had a casino we could put a large sign outside it that said, 'this casino is funding the schools of Brčko'.
>
> AJ: But Brčko already has a casino in the boat moored on the River Sava.
>
> DM: [smiles] Shhh! That one is illegal.
>
> (Excerpt from an interview with the Brčko District Mayor, Brčko, 8 May 2003)

Though the Supervisor did not confer legality on the boat casino, neither was there the political will to remove it. We see in this porous relationship between legal and illegal an example of what James Holston (2008: 19) refers to as the 'misrule of law' when describing a 'system of stratagem and bureaucratic complication deployed by both state and subject to obfuscate problems, neutralize opponents, and, above all, legalize the illegal'. While the ability to categorize certain acts as legal or illegal is often recognized as a characteristic of state power (Engler, 2003), in this case it is the capacity to suspend recognition of which category the boat belongs to which emphasizes the Mayor's authority.

This tenuous nature of the legal/illegal boundary was evident in other areas of economic reform in Brčko, such as the regulation and privatization of public housing. The demand for housing in Brčko has far outstripped supply since the conflict on account of the large number of Serb displaced persons in the town and the relatively high levels of return. To meet this demand extra floors had been built illegally on top of existing residential buildings. This was described by a UN official in Brčko as 'a neat idea'[25] to cater for the increased population in the town. Like the casino boat there was no mention of these extra floors in official documentation such as Court lists or OHR documents, and they were neither regulated nor was it clear who owned them. While this was proving a serious barrier to privatization it was not until there was a serious fire in April 2003 that the Supervisor intervened to ensure that 'all issues related to public buildings are clearly regulated before the process of privatization of the apartments is concluded' (OHR, 2003; see Figure 6). Again, it was through the deployment of a Supervisory

Figure 6   Burnt apartment building, Brčko, April 2003
Source: Author's photograph

Order that the apartments were rendered legal and visible to the law (through inspections and regulation) while also functioning as a catalyst to the privatization process.

The most controversial example of the production of illegality and the associated incorporation of economic activity into the gaze of the improvised state is the case of Arizona Market. By 2002, the Market had grown to over 2500 stalls, sprawling next to the main road between Brčko and Tuzla. The Market was Janus-faced: it was a multi-ethnic goods and produce trading area with a regular cattle market representing 'an engine of peace' (Feit and Morfit, 2002), and at the same time a zone of human trafficking, drugs smuggling and arms trading (see Sherwell, 2000; Voss, 2001). The former sentiment is further evidence of the primacy given to mathematical multi-ethnicity in Brčko, as any activity, regardless of legality, is given a positive sheen by the existence of people from 'different ethnicities' carrying it out. From 1999 onwards, preparations were made to privatize Arizona Market. Following an EU and UNICEF co-initiative in 2001 entitled 'STOP', the brothels and trafficking way-stations at Arizona Market were shut down, representing another manifestation of the improvised nature of the state in Brčko District (UNICEF, 2002: 65). The DMT formulated a concept paper detailing a strategy to legitimize trade at Arizona and register all of the stallholders, directed towards encouraging private investment (Feit and Morfit, 2002). This seems a particularly audacious proposal, as Arizona Market could be seen as the most privatized space imaginable, with little or no state intervention or regulation. The idea of attempting to encourage further private investment demonstrates the symbolic capital of legitimizing certain investment. One of the key difficulties behind the proposed privatization of Arizona Market was purchasing the land, rumoured to belong to a Croatian 'mafia group'.[26] This was resolved through a compulsory purchase order in 2002, bringing this site under the gaze of legal regulation and legislation for the first time in ten years. Following a bidding process the site and market were sold to *ItalProjekt*, an Italian–Bosnian joint venture, which held plans to redevelop the market as a large shopping complex, with all the stores in uniform warehouses.

This process of privatization rendered Arizona Market visible to the instruments of taxation and regulation, as each stall was relocated to space inside prefabricated warehouses. I expected this to lead to protests among the stallholders, but when I discussed this issue they seemed uninterested. One stallholder typified this attitude by commenting: 'really, we pay our money to the Croats or to the people from *ItalProjekt*, it makes no difference to us'.[27] This comment provides an insight into the interchangeable nature of the 'state' in a system of uncertainty and improvisation. The stallholder was referring to paying for the stall in order to guarantee security, look after the interests of the market and pay for external lighting. Arizona Market was far from anarchy prior to 'regulation'; it was similarly regulated, but only by instruments that were invisible to the gaze of the state. This scenario of institutionalization and regulation conjures images of Nordstrom's 'non-state sovereignty', used to describe the governance exercised by networks of 'informal' actors (Nordstrom, 2003: 340). While privatization has not removed these patterns of regulation it has, at least partially, brought them into the machinations of the improvised state in Brčko District.

These examples demonstrate that the economic management of Brčko should not be conceived as a technical matter, divorced from more 'political issues'. Since the Final Award, the rendering of economic practice as privatization has revitalized the authority of the District Supervisor and strengthened the agencies of the improvised state as they have been granted the power to legitimize certain practices. These practices of state regulation have operated through a variety of techniques and tactics, including passing laws, Supervisory Orders and concept papers. At the same time, the District Government has been weakened as assets have been sold and 'new' discourses of privatization have been endorsed. These processes of revitalization of the improvised state are also reflected in the management of the symbolic landscape in Brčko District.

4.4   Neutralization: Making People Think

As discussed above, the greatest challenge of implementing the Final Award identified by intervening agencies was 'making people think in terms of Brčko District'.[28] In light of the urban destruction committed during the conflict, an attempt to create a coherent social container out of the fragmented landscape of Brčko District was an extremely difficult undertaking. Following the harmonization of the institutions of government and fiscal regimes, the task for the Assembly and OHR was to ensure that the symbols of Brčko District were 'politically and ethnically neutral' (Final Award in OHR, 2000: 286). The unification of the symbols of the District is a vital aspect of reducing the 'micro-level humiliation and contempt' (Dahlman and Ó Tuathail, 2005b: 21) suffered by returning refugees and DPs. In addition to encouraging refugee return, an examination of practices such as relabelling houses, renaming the streets and shaping public commemoration draws attention to the important role that reintegrating the symbols of the District has played in building local state capacity.

The process of symbolic 'neutralization' was begun before the creation of the District with the allocation of new identity cards to Brčko residents, without the Serb insignia of the twin-headed eagle (see above). The Final Award increased the pace of neutralization by declaring three official languages across the District (Serbian, Bosnian and Croatian)[29] and two official scripts (Serb Cyrillic script and Bosniak and Croat Latinic script). Since this equalization applied to all public signage, Brčko District became, in 2000, the only place in BiH to have road signs in both Cyrillic and Latinic script (see Figure 7).

Following the equalization of Serbian and Cyrillic scripts, the Assembly passed a resolution declaring the renaming of Brčko's streets to remove the names given during the Serb occupation in 1992. As new street names were allocated, each building was given a uniform yellow and blue house number-plate displaying the name of the street in both Cyrillic and Latinic scripts (see Figure 8). This colour scheme is derived from the Bosnian flag imposed by High Representative Carlos Westendorp in February 1998 (see OHR, 1998 and Ó Tuathail, 2005b). The new street names aimed to conjure a shared Yugoslav past; hence *'Bulevar Đenerala Draže Mihajlovća'* ('Boulevard General Draža Mihajlović') was changed to *'Bulevar Mira'* ('Boulevard of Peace') and *'Srspskih Oslobodilaca Brčkog'* ('the Serb Liberation of Brčko') to *'Bosne Srebrene'* ('Silver Bosnia'). Specific strategic objectives were also written out of the landscape, as the road to the Croatian Krajina (*'Krajiški Put'*) was renamed *'Dejtonska'* (after the GFAP). This example demonstrates the shift in power relations following the creation of Brčko District: the vision of 'greater Serbia' (seen as dependent upon a

Figure 7 Dual script road sign, Klanac, Brčko District, February 2003
Source: Author's photograph

Figure 8 New house number, Brčko, July 2003
Source: Author's photograph

connection with the Croatian Krajina region) was literally erased, only to be reinscribed with the new vision for BiH, that of the GFAP. This procedure seems to exemplify the assertion of Azaryahu (1996) that the act of renaming 'asserts that a radical restructuring of power relations in society has indeed been accomplished, or is underway, and it indicates a profound reconstruction of social and political institutions' (Azaryahu, 1996: 318).

There are, however, spaces where the uniformity of Brčko District's symbolic landscape has been disrupted, as nationalist scripts and alterations have subverted the new street signs, house number-plates and place names. One tactic used across the District was to obscure with paint the script of other nationalities; thus Bosniak and Croat villages or houses erased the Cyrillic script while Serbian villages erased the Latinic script. These practices gave a visceral illustration of the struggle between new state institutions and those who wished to protest against the renaming process, as the District authorities would repeatedly clean place-name signs only for paint to reappear over the Cyrillic or Latinic script. In addition, graffiti covers many of the buildings, bridges and walls in Brčko, including 'tags' of particular graffiti artists, rap and rock band names, as well as nationalist messages. In a series of places across the District the word '*ЛУКА*' ('Luka') has been daubed on buildings, in Cyrillic to situate its national affiliation, to act as a haunting reminder of the wartime atrocities committed at Brčko port. Other anti-OHR messages are often communicated in the graffiti, for example a wall near the centre of town had been daubed with the words '*JEБЕM BAM ДИСТРИКТ*' ('Fuck your District'). In addition to the use of Cyrillic script, it is instructive that the word used for the District is not *opština*, but rather '*Distrikt*', a localized spelling of the English word used to denote the new political entity created through the arbitration process. This use of graffiti was not restricted to the urban areas of the District; Serb farms whose owners had been evicted were often daubed with an Islamic crescent insignia as a way of taunting returnee families (see Figure 9).

Other examples from urban Brčko demonstrate the learnt nationalisms within contemporary Brčko District, as antagonistic political slogans have been used to intimidate returnees and claim urban territory. In an example from central Brčko a graffiti artist used Cyrillic to write '*БОГ И РАДОВАН*' ('God and Radovan'). This comment suggests a battle cry, with the reference to the Bosnian Serb leader and founder of the SDS, Radovan Karadžić. The interesting point in this example is that the Cyrillic has been corrected, as the author originally wrote a Latinic 'D' as opposed to Cyrillic 'Д' (see the top picture of Figure 10). A similar mistake can be seen in the picture at the bottom of Figure 10, where the graffiti simply reads '*СРБIУА*' ('Serbia'), though again the author has corrected their Cyrillic, mistakenly writing a Latinic 'I' instead of a Cyrillic 'И'. One plausible reading of these error-strewn examples is that they display the urgency to embrace linguistic difference in Brčko; Cyrillic has been learnt in order to embrace its nation-defining qualities.

Figure 9   Farm well in Bukvik, Brčko District, June 2003
Source: Author's photograph

It is not my intention here to engage in a textual analysis of Brčko's graffiti – rather to demonstrate how the obscuring of road signs and the use of nationalist slogans has interrupted and unsettled the mechanisms deployed by the new state institutions to reorder the symbolic landscape in the towns and villages of Brčko District. While 'neutral' place names have reversed the practices of nationalist naming during the conflict, the strategic placement of graffiti has allowed the District to continue to be haunted by the violence and division of its past. These practices seem to echo Cresswell's (1996) contention that graffiti acts as a 'tactic' of the dispossessed – 'a mobile and temporary set of meanings that insert themselves into the interstices of formal spatial structure' (Cresswell, 1996: 47; see also Ross *et al.*, 2001). In doing so the connections between territory and identity are (temporarily) reinscribed, as sites within the urban and rural areas are reconnected with particular nationalist projects.

This process of reconnection is also evident in the commemoration to the 1992–95 conflict. During the period of the fieldwork, the Supervisor had used the absence of a specific mandate in the Final Award to 'avoid most cultural, historical and purely national issues altogether' (Clarke, 2004). This attitude was also reflected in the responses from other OHR Officials who described memorialization as a 'soft issue'.[30] However, in place of a cross-national memorial to acknowledge the events of the conflict, a range of monuments stood as testament to specific mythologized historical narratives,

Figure 10   Examples of graffiti, Brčko town centre, December 2002
Source: Author's photographs

such as the statue of Draža Mihajlović and the 'Serb Defenders of Brčko' (see above). These examples became rallying points for the Serb community across Brčko District, sites where sporting victories are celebrated, wedding photos taken and the Battle of Kosovo marked with the laying of wreaths and the reading of Serb epic poetry. Efforts by the OHR to remove these monuments had been limited, acknowledging that such actions would 'stir nationalist sentiments'.[31] Despite this, following lobbying by Bosniak returnee groups and with the assistance of the OHR, in 2004 the statue of Draža Mihajlović was moved to an Orthodox Cemetery on the outskirts of Brčko.

The partial commemoration in Brčko guides the popular consciousness away from remembering the events of genocide and 'ethnic cleansing' that scarred the town in 1992 and dramatically changed its demographic constitution. While the street names stimulate a communal memory of Yugoslav heroes such as Ivo Andrić or Nikola Tesla, collective commemoration of the conflict itself had not been attempted. As will be discussed in Chapter Six, the forms of transitional justice enacted at the ICTY have failed to cultivate local engagement in trial proceedings, while allowing nationalist politicians to question the neutrality of the Court. Echoing Tito's approach to inter-ethnic reconciliation following the Second World War, through the absence of shared commemoration of the events between 1992 and 1995 the Brčko community has been encouraged to collectively forget (Judah, 2000: 136). While the neutralization of the symbolic landscape in Brčko has served to strengthen state capacity by presenting a unified geography, its silence on issues of war crimes and culpability has permitted the continued recycling of fractured and mythical histories among nationalist political parties.

4.5 Conclusion

International intervention in BiH has often been framed as a measurable exercise, a multifaceted policy of strengthening institutions of governance, installing democracy, ensuring economic prosperity and encouraging inter-ethnic reconciliation (see, for example, UNDP 2002, 2009). This reassuring teleology from dependence to independence is reflected in a series of measurable indicators, used to gauge the stage which BiH has reached, such as the holding of democratic elections, the 'multi-ethnicization' of the police force and the withdrawal of international troops. Media and academic texts often reduce these indicators to the simple shorthand of an 'exit strategy', a time when international involvement has ceased and independence has been achieved (Rose, 1998). Through an examination of the exercise and durability of international intervention in BiH this chapter has questioned such teleological narratives.

Rather than the gradual centralization of power in new state institutions, this chapter has illustrated the more complex geography of state building in Brčko District. This discussion has illuminated three significant binaries that shape the improvisation of the state. The first concerns temporality, where state institutions attempt to perform permanence through ephemeral acts. This tension is perhaps most explicit in the creation of Brčko District Day, where an invented tradition seeks to celebrate a new form of collective (and territorial) identity in a similar fashion to Tito's recourse to 'brotherhood and unity' (see Chapter Three). But claims to permanence are also conveyed through the range of supervisory powers used to establish new governmental institutions. Discursively, these claims are reflected in the attachment of intervening agencies to a language of 'stability,' where the establishment of the District is seen as a form of territorial fix that addresses competing ethnic claims. These improvisations draw on the resources laid down first at the GFAP negotiations and later through the Brčko arbitration process. As the Supervisor says, the Final Award granted the mandate to reorganize processes and institutions of governance. One of the important conclusions regarding this process is the limits to these powers, since the OHR was dependent upon a range of other international and local institutions in order to perform such claims to statehood.

The second binary that is illustrated across the examples of this chapter relates to the relationship between multi-ethnicity and the continued use of ethnic categories in order to establish new state practices. While Brčko constitutes a rare challenge to the ethnopolitics endorsed at the GFAP, the continued use of ethnic categories and equations in the establishment of government institutions points to a narrow understanding of multi-ethnicity as the incorporation of multiple ethnic groups in a single institution. This understanding of multi-ethnicity was illustrated in the delicate ethnic balance of new appointments to the District Government and Assembly and the continued reliance on the Yugoslav census as the ultimate template of legitimate ethnic weightings within any given territory in BiH. The production of a new symbolic landscape for Brčko similarly illustrated this interpretation of multi-ethnicity, where practices of ethnic 'neutrality' emphasized coexistence of different ethnic groups rather than forms of social integration. Of course, such performances of multi-ethnicity enabled higher levels of refugee return and were, perhaps, the only possible strategy for creating a multi-ethnic territory in the wake of the GFAP. But it is important to acknowledge that such practices do not challenge the basis of ethnopolitics, structured as it is around assumptions of the impossibility of plural ethnic identities. Anthropological work by Tone Bringa (1995) and Stef Jansen (2007) has illustrated a more nuanced nature of identity in BiH, as everyday collective practices challenge the exclusionary ethnic categories. The challenge posed by incorporating a more plural understanding of identity into new governmental practices points to the value of ethnic categories as symbolic resources for intervening agencies in BiH. Emphasizing the theoretical framework set out in Chapter Two, this chapter has shown how the OHR has performed new state practices in Brčko by using the resources granted by ethnic assumptions made at the GFAP.

The final binary that shaped the improvisation of the state in Brčko is between authoritarianism and democratization. The discursive challenge for intervening authorities has been to present practices of intervention as democratizing, while suspending opportunities to vote and thereby participate formally in the political process. Indeed, even if electoral competition had not been suspended in the District this would not grant the opportunity for individuals to influence the priorities of the OHR. In order to facilitate democratic politics the Supervisor emphasized the significance of associative life. The cultivation of civil society was deemed by the OHR and other international organizations to be evidence of fledgling forms of political participation. Such participation was conceived in theatrical terms by the District Supervisor: that civil society groups could constitute an audience to the performances of government undertaken within District institutions. One of the challenges to this vision of political participation was the reticence of civil society groups to perform their role as audience, since it was unclear what was at stake in District Government decision making. These groups understood that the decisions ratified by the District Government could be overruled by the Supervisor if they did not accord with the priorities of the Final Award. We see in these unmet expectations of international agencies the significance of understanding the state as a practice, as the role of institutions shifts between planning and implementation.

The following chapter explores in more detail the relationship between democratization and authoritarianism in the improvisation of the state in BiH. In particular, it examines how intervening agencies have attempted to cultivate civil society organizations in the post-GFAP period. This discussion focuses on the processes through which international and domestic agencies have conflated the existence of NGOs with a thriving civil society. This approach reflects the longer historical trajectory explored in Chapter Three, where the framing of the violence in BiH as a 'humanitarian disaster' ensured that NGOs and other non-state agencies became important actors in the international response. As detailed in the discussion above, the positioning of NGOs as evidence of civil society is particularly important in the case of Brčko District, where the suspension of electoral competition only emphasized the significance of associative life in the practice of democratic politics. It is argued that the placement of NGOs in these terms allows emerging state agencies to perform their verticality, where they are conceptually separated from society. But the evidence from the practice of intervention suggests a considerable blurring between different articulations of the state in BiH.

Notes

1   Interview with OHR official, Sarajevo, 2003.
2   *Mjesne zajednice* (MZs) were introduced in the 1974 Yugoslav Constitution as the lowest strata of socialist government, enabling citizens to participate directly in the management of society (Singleton, 1976: 312). MZs are arranged territorially, they traditionally had no inherent ethno-national affiliation (at least in urban areas) and, reflecting the patriarchal nature of Bosnian society, their participants are primarily men (Bojičić-Dzelilović, 2003: 296; Dahlman and Ó Tuathail, 2005b; Pusić, 1975). Though MZs have lost official government competences in post-conflict Brčko on account of the 2003 Law of Local

Communities of Brčko District of Bosnia and Herzegovina, they continue to act as important sites of communal participation and have been crucial within the returns process (see Chapter Five).

3 See in particular Jusuf Kadrić (1998) *Brčko: Genocide and Testimony*, Human Rights Watch (1992) *War Crimes in Bosnia-Hercegovina*, and indictment documents and witness statements at www.un.org/icty/indictment/english/ jel-2ai981019e.htm and www.un.org/icty/indictment/english/ces-3ai021126e. htm. For an account of the life of Goran Jelisić see Drankulić (2004).

4 Interview with former OHR official, 20 February 2003.

5 The UN-led International Police Task Force.

6 Referred to as simply 'the Supervisor' for the remainder of the chapter.

7 During a meeting of Croat and Serb forces in 1994, Croat forces asked for control of Brčko. Radovan Karadžić erroneously reported that 'the Serbs refused because they built it and it was completely Serb' (Silber and Little, 1996: 308).

8 Interview with Ambassador Clarke, Brčko, 24 March 2003.

9 Interview with former Brčko OHR official, Brčko, 19 May 2003.

10 This aspect of Supervisory power was discussed by Ambassador Clarke, Brčko, 24 March 2003, an OHR Democratization Officer, Brčko, 24 May 2003, and an OHR official, Brčko, 2 June 2003.

11 Interview with Ambassador Clarke, Brčko, 24 April 2003.

12 Over this period, residents of Brčko municipality continued to vote in state-level presidential elections.

13 Interview with OHR official, Brčko, 24 May 2003.

14 In contrast to the Serb SDS, the Croat HDZ and the Bosniak SDA, the SDP is committed to a multi-ethnic Bosnia (and consequently benefits from a strong Bosniak backing) and is well represented in industrial areas and among the military (see European Forum, 1999).

15 Bosnia has yet to establish a multi-ethnic state police force as the RS have rejected the creation of Local Police Authorities (LPAs) because their borders crossed the IEBL (see Wisler, 2007).

16 Interview with the Brčko Mayor, 8 May 2003. Ironically, this Mayor was replaced in 2004 following an indictment for abuse of office connected to constructing extra floors on government-owned buildings.

17 Interview with Ambassador Clarke, Brčko, 24 April 2003.

18 The lack of confrontation within the District Government was discussed by a former OHR official, Brčko, 14 May 2003, Ambassador Clarke, Brčko, 24 April 2003, and a representative from the SDS, Brčko, 14 April 2003.

19 Interview with OHR official, Brčko, 2 June 2003.

20 Interview with SDS representative, Brčko, 14th April 2003.

21 Interview with SDS representative, Brčko, 14 April 2003.

22 Interview with Brčko representative for the National Democratic Institute, Brčko, 3 June 2003.

23 Interview with environmental NGO worker, Brčko, 15 May 2003.

24 The line for 'general poverty' was calculated as on or below 1843KM (approximately £615) per year (UNDP, 2002: 50).

25 Programme Manager, UNDP Brčko Local Action Programme, Brčko, 18 February 2003.

26 Though it is difficult to verify whether Croatian criminal groups owned the land on which Arizona Market was situated, ownership was framed in these terms during interviews with representatives from Brčko NGOs and international organizations.
27 Interview with Arizona Market stallholder, Brčko, 3 May 2003.
28 Interview with former OHR official, Brčko, 20 February 2003.
29 The grammatical differences between these languages are only slight, though Croatian differs in many of its nouns from Serbian and Bosnian (for the politicized nature of Bosnian linguistics see Stankovic, 2000: 59 and Holbrooke, 1999: 232).
30 Interview with OHR official, Brčko, 24 May 2003.
31 Interview with OHR official, Brčko, 24 May 2003.

Chapter Five

Gentrifying Civil Society

I'm reading Jan Morris's (2002) *Trieste: The Meaning of Nowhere*. I have noticed that travel writers spend a great deal of time talking about coffee, including accounts of drinking coffee, the production of coffee or the domestic rituals that surround coffee. It reminds me of the social science work that explores the role of coffee, and coffee houses in particular, in the growth of civil society in Europe in the 19th Century. It is the same in Morris: she discusses the role of the Austro-Hungarian coffee houses in Trieste as centres of exchange: both of goods and ideas. They were mixing points under the ornate rococo ceilings. This must have been the case in Brčko once, with the Grand Hotel in the centre of town, fondly remembered by older residents as a place where people would meet to drink on the open veranda in the centre of town. It is not so now, where there are clear distinctions made by interviewees between 'our cafes' [*naši kafići*] and those of other ethno-national groups. The distinctions are subtle, which television channel is being shown, which football shirts are on the wall or in [which] language your receipt is printed. In these instances, coffee in Brčko is no longer the catalyst to mixing but becomes instead another instrument of social segregation.

Field journal, 15 September 2004

The idea of civil society is elusive. There is a tendency among scholars to use the term to refer to either a concept or a practice. The concept of civil society is suggestive of a 'realm' of institutions and individuals 'between' the state, the market and (sometimes) the family. This relational understanding establishes a conceptual distance between civil society and the state. In

The Improvised State: Sovereignty, Performance and Agency in Dayton Bosnia, First Edition. Alex Jeffrey.
© 2013 John Wiley & Sons, Ltd. Published 2013 by John Wiley & Sons, Ltd.

contrast, understanding civil society as a practice suggests a rather more enmeshed institutional landscape. The dividing lines between state, market and family become blurred, as powerful agents seek to shape the practices of notionally autonomous institutions. But underlying this distinction is a lingering normative impulse, characterized by my words in the field journal above, that civil society is a social good and reflective of (and necessary for) democratic politics. Through this lens civil society is not just a means of participation in the state; it is a mechanism whereby that participation is visible to other members of a political community. Civil society is thus conceived as a site of the political, a public arena of contestation and differ-ence negotiated through non-violent interactions. In this sense civil society resembles the social consequence of public space. But, as intimated in the field journal, the presence of public spaces does not necessarily lead to the forms of interaction and civil contestation imagined within concepts of civil society. Forms of social, cultural and economic stratification lead to the fragmentation of actual existing public spaces, which may challenge their imagined virtuous role.

The numerous layers of institutions and practices of civil society are lost in some of the discourses of post-Cold War state building. Partly this is due to the complex temporality in the forms of political transformation that international agencies are encouraging. The prominence of civil society as a realm of political participation emerged following the transitions to market democracies in Eastern and Central Europe in 1989 and into the 1990s. Agencies intervening in the post-Yugoslav states held these move-ments as illustrations of the capability to shift from one-party Communist politics to a more plural institutional landscape where numerous political affiliations and desires could be accommodated. But, of course, one of the central attributes of the movements in Hungary and Poland was their autonomy: they emerged in opposition to political elites and often sought to claim the power of the state as their own. While perhaps cultivated in the coffee shops, public squares and streets of Warsaw or Budapest, these move-ments went on to lay claim to formal political power as they crossed the imagined line between civil society and the state. This makes the centrality of cultivating civil society in state-building situations more remarkable, as state elites seek to cultivate forms of imagined political dissent. As we will see in this chapter, these actions often constitute attempts to create political alterity, which remains within the authority of the improvised state.

While these debates relate to fundamental questions regarding associative life, political participation and social transformation, discourses of post-Cold War state building take a more institutional approach to the existence and significance of civil society. In this sense, attempts to cultivate civil society have focused on nurturing NGOs run by volunteers and directed towards social welfare objectives and reconciliation between previously antagonistic groups. This dual function (welfare and reconciliation) immediately places NGOs in a contradictory political space: they are both part of the state in terms of the delivery of services, and (and often within the same organization) supposed to provide the opportunity for people to come together to resolve contestation with civility. This precarious institutional positioning creates ambiguity among the wider population as to the purpose or role of such institutions within the new institutional architecture of emerging states.

This chapter explores this ambiguity in the case of state building in Brčko District. The discussion centres on the ways in which intervening agencies, predominantly the Office of the High Representative (OHR) and the Organization for Security and Co-operation in Europe (OSCE), have conflated NGOs with the existence of a thriving civil society. As indicated by the previous chapters, this is particularly important in the case of Brčko, where the suspension of elections meant that a vibrant civil society could be used by the OHR to indicate democratic participation. Consequently, new forms of funding and institutional support were established by the OHR and donor agencies in order to stimulate NGO activity. It is argued that these practices have conferred legitimacy on particular political practices and cultural styles, while sidelining or legislating against other practices. It is through this discussion that the resourcefulness at the heart of state improvisation is brought to the fore, as regulating bodies draw on the legitimacy of 'civil society' in order to establish an arena of compliant organizations. In turn, newly formed NGOs have exploited a range of repertoires, in order to gain respect and access to funding, from improving English language skills to developing links with hard-line nationalist political parties such as the *Srpska demokratska stranka* (Serb Democratic Party or SDS). This discussion highlights the contradictory nature of state improvisation: while it provides space to challenge the often anti-democratic nature of international intervention, it illustrates that the resulting critiques are not necessarily emancipatory. They are instead offering differing visions of state performance, enshrined in particular understandings of identity and territory based on ethno-national affiliation.

The chapter is divided into four sections. In the first I provide an overview of the political and intellectual background to recent studies of civil society. In particular I examine theoretical approaches that have sought to challenge the image of civil society as an autonomous sphere, exploring instead the symbolic power of imagined autonomy. I argue that the power of civil society comes from its imagined separation from the state, while simultaneously the state retains a series of legislative, economic and cultural instruments through which its will may be expressed in civil society. In order to narrate these relationships I engage again with the work of Pierre Bourdieu, drawing on his metaphor of social and cultural capital to illustrate the relationship between and among state organizations and civil society in BiH. Sections 5.2 and 5.3 examine different aspects of Bourdieu's schema. In section 5.2

I explore attempts by NGOs in Brčko to accumulate social capital, a set of examples that illustrates the highly individualized struggles undertaken by these institutions to build trust by establishing political contacts. I argue that as NGOs have professionalized and attempted to represent 'civil society' they have been drawn into the gaze of international organizations and the regulatory instruments of the local state. Rather than acting as sites of communal participation, as frequently suggested within discourses of civil society, this evidence suggests that NGOs have engaged in a struggle to accumulate the necessary social and cultural resources to receive donor funding and access institutions of the local state. Section 5.3 develops this argument, exploring how NGOs sought to portray 'desirable' cultural traits, such as language skills or knowledge of donor discourses. But, crucially, this discussion illustrates the plural centres of power that seek to improvise the state in BiH, as some institutions sought to secure funding from nationalist political parties that sought to challenge the authority of the District Government.

The processes outlined in sections 5.2 and 5.3 have left a group of 'gentrified' organizations that had accumulated the necessary social and cultural capital to negotiate the barriers to recognition as legitimate civil society agents. But these are not, of course, the limits of associative life in the District. There are numerous organizations that are not recognized by intervening agencies as legitimate representatives of 'civil society' and are thereby cast out as either illegal or lacking legitimacy. In the final section I examine one set of organizations that falls into this category: the *mjesne zajednice* (MZs). These territorially defined community groups, or wards, were established as the lowest strata of government in the 1974 Yugoslav constitution, performing sub-municipal governmental functions such as the assessment of infrastructural needs and the collection of taxation (see Pusić, 1975). Until 2004, these associations had no legal basis within Brčko, amid concern over the subdivision of the authority of the District Government. This situation contrasts with the two Entities, where local government legislation in 1995 (Federation) and 2005 (RS) has continued to specify a role for MZs in the governmental process (see Bajrović, 2005). In the postwar period many of these organizations transformed into advocacy groups for the return of displaced people, especially where MZs formed 'in exile' from their original locality. In the case of Brčko there was considerable uncertainty among intervening agencies as to the role of MZs in the governance of the District. While they performed an important function as sites of community deliberation and trust, they were denied the possibility of applying for funding or project work on account of their formerly 'governmental' status. In the District the levels of activity of MZs varied; their leadership was often also involved in formal political parties and participation was almost exclusively male. Thus, these were organizations that could not be easily categorized: they challenged the image of civil society as non-party-political, they did not organize their work into distinct 'projects', and they were not representative of wider Bosnian society. But the imperfections of the MZs did not diminish their role as centres of trust and deliberation for Brčko citizenry. In some sense, their continued existence pointed to the limitations of the gentrified civil society organizations to perform the functions expected of them.

5.1   Building Civil Society

[I]nstitutions that have been created by imposition will never function effectively unless Bosnians of all ethnicities buy into them and until Bosnian citizens expect them, and not international organisations, to deliver reform, exercise democratic rules and procedures day by day in a bottom-up process of building the state.

Schwarz-Schilling (2006: 84)

The comments by the then High Representative Christian Schwarz-Schilling reflect one of the central tensions in state building in BiH. The establishment of new laws and institutions has been a markedly top-down affair, enabled by the Bonn Powers retained by the High Representative. But despite these practices, the OHR and other international agencies have continued to celebrate the significance of 'bottom-up' processes, involving an active citizenry that shapes a new politics. As Partha Chatterjee (2004: 27) has indicated, this claim to popular consent has been at the heart of modern understandings of the state since the French Revolution. Chatterjee suggests that the legitimacy of the state is grounded in an assumption of popular sovereignty: that rule is being conducted through the consent of the citizenry. This tension between imposition and popular sovereignty has required a set of discursive tactics that simultaneously enshrine international obligation to supervise state institutions while attempting to cultivate compliant forms of popular participation. Compliance is significant here, since, as we have seen, there are alternative ideas of what constitutes legitimate sovereignty in BiH, often structured around imagined connections between ethno-national identity and statehood (see Campbell, 1998a). The quest to establish Dayton BiH as the legitimate sovereign form has therefore required a parallel process of cultivating forms of popular political participation and consent for new forms of rule. These processes have been gathered under the term 'building civil society'.

The political antecedents to the promotion of civil society can be traced to two, roughly contemporaneous, political transformations. First, and as discussed above, policy makers have drawn inspiration from the role played by dissident and pro-democracy groups in the fall of Communism across Central and Eastern Europe in the later 1980s. These movements have been portrayed by both scholars and Western policy analysts as a victory for civil society: the will of the people holding the state to account (Bernhard, 1996; Crook, 2001; Offe, 2004). But the universal celebration of civil society that has characterized some of the subsequent commentaries not only denies the historical specificity of the events in question, but also erases the significant struggles and anti-democratic behaviour which occurred within and between movements. Second, discourses of international development over the 1980s and early 1990s began to promote economic and social development through the non-state actors, in particular NGOs. This neoliberal ideology was encapsulated by World Bank initiatives structured around fostering 'good governance' and International Monetary Fund Structural Adjustment Policies. These initiatives, while comprising considerable variation in terms of tactics and outcomes, sought to question the state as the focal point of developmental interventions.

The political roots of the promotion of civil society have longer intellectual lineages. The origins of the popularity of civil society can be found in the work of Socrates, Aristotle and Plato, where the term described the use of rational argument to resolve conflict and achieve a common good. These early models of civil society reflect the uses of civil society as a mechanism for mediating between '[...] selfish goals of individual actors and the need for some basic solidarity' (Hann, 1996: 4). The scholars of the Scottish Enlightenment, Adam Ferguson, Adam Smith and David Hume, developed these themes in the eighteenth century. These scholars conceived civil society as not only a social or political consideration but also an economic practice, involving the civilized exchange of both ideas and goods. This raised the question of the role of the state: if economic transactions could be conducted with civility, the intervention of the state was seen, particularly by Adam Smith, as unnecessary.

In the nineteenth century, Hegel was more circumspect in reducing the role of the state in managing economic and social affairs. Hegel saw civil society as a consequence of the capitalist mode of production which constituted a location of competing class relations, each with little interest in the common good. Consequently, the state was vital in mediating the 'egotistical tendencies' of certain elements of civil society (Padgett, 1999: 4). This moral role of the state was later contested by Marx, who felt that the private dimension of civil society overpowered the public aspect, which, in a capitalist society, resulted in an overemphasis on the rights of the individual to pursue self-interest (Arato and Cohen, 1995: 142). Despite their differences, these models introduce the notion of civil society as external to the state, potentially holding the state to account. In addition to a philosophical aspiration or a theatre of economic exchange, Marx and Hegel conceived civil society as a focal point of democratic practice. This role was emphasized by Alexis de Tocqueville in *De la Démocratie en Amérique* (Democracy in America) (1835). In this work, de Tocqueville argued that the guarantee of individual liberties was to be found in what he called 'democratic expedients', which included local self-government, the separation of the church and the state, a free press, indirect elections, an independent judiciary and, above all, 'associational life' (Kaldor, 2003: 19).

The conflation of civil society with associational life has been evident in the rhetoric of intervening agencies in BiH. In particular, there has been a desire to portray civil society as a democratic prerequisite, separate from the state. For example, in April 2011 Jasminka Džumhur, the Ombudsperson of Bosnia and Herzegovina, gave a speech at the OSCE that defined civil society thus: 'Civil society is a sphere of institutions, organisations, networks and individuals, located between family, state and market, and citizens of their own will establish associations for advocating of their common interests. In its simplest form civil society is a group of institutions and associations/organisations, linking citizens with governments and the private sector' (Džumhur, 2011: 1; see also OHR, 2006). Civil society is thus represented as a space of liberty, a democratic expedient in de Tocquevillian terms, capable of holding the state to account and negotiating individual and collective needs. This representative strategy allows NGOs to be equated with 'civil society' on account of their perceived autonomy, coupled with their supposed ability to 'pluralize the institutional arena and bring more democratic actors into the political sphere' (Mercer, 2002: 10). Such a perception has led to the suggestion that their mere presence in post-conflict BiH is evidence of democratization (Tvedt, 2002: 364). This line of thinking reflects quantitative ratings of democratization, where the number of NGOs correlates with the level, or 'depth', of democracy (Kaldor and Vejvoda, 1997: 77; Mercer, 2002: 10). Perhaps the most stark examples of such an instrumental approach are contained in the work of organizations such as Civicus or the Open Society Institute as they attempt to construct 'indices of democracy' largely based upon numerical analyses of NGO presence (Anheier, 2004).

The image of separate spheres of state and civil society is a powerful one. It has become part of the lexicon of state building and developmental interventions, a means through which autonomy, popular participation and democratic validation may be conveyed (Fagan, 2010; McIlwaine, 1998). But such crisp social and political boundaries do not stand up to empirical or theoretical scrutiny. The imagination of state and civil society as distinct spheres neglects to acknowledge the nature of the state as a coercive system that has capacity to intervene in the constitution of civil society. The line between state and society is further blurred by a focus on improvised state practices, where the state is understood to come into view through a range of social practices dispersed through 'public' and 'private' space. If we look at the work of critical state theorists a more nuanced understanding of civil society comes into view. Navaro-Yashin's (2002) study of the production of ideas of civil society in Turkey is particularly helpful in this regard. Following an ethnographic examination of the idea of the state in Turkey, Navaro-Yashin suggests that those who celebrate a new associational pluralism 'outside' the state in Turkey may have confused a changing discourse or technique of state power with an autonomous rise of civil society (Navaro-Yashin, 2002: 132). In this way, Navaro-Yashin suggests that there is no 'autonomization' to be observed, but rather a changing enmeshed relationship (ibid.).

At the forefront of Navaro-Yashin's ethonography is an engagement with Timothy Mitchell's concept of the 'state effect' (1999). As discussed in Chapter One, Mitchell draws on Foucault's idea of 'discipline' to examine the *productive* nature of state rule, a form of power which works not from the outside 'but from within, not at the level of an entire society but at the level of detail, and not by constraining individuals and actions but by producing them' (Mitchell, 1999: 86). This account is helpful in two ways. First, it foregrounds the enshrinement of state within the details and patterns of social life. While the idea of the state is suggestive of an abstract political entity, it is summoned into existence through everyday interactions and practices. This conceptualization of the state is close to a concept of improvisation, though perhaps with less emphasis on the forms of tactics and strategy that shape these moments of state performance. Second, it helps illustrate the ways in which the imagined distinction between state and civil society is politically significant. For Navaro-Yashin, referring to state–civil society relations strengthens the idea 'of a state that is self-contained and does not meddle, more than its due, in the affairs of society' (Navaro-Yashin, 2002: 136). The ability to discern civil society thus becomes a key mechanism through which state power is legitimized.

The political effects of the creation and maintenance of a state–society distinction extend beyond the reproduction of certain forms of state power. If we use Bourdieu's conceptual vocabulary, we can see how the state agencies operating in BiH have placed symbolic capital on certain traits of 'civil society' while legislating against others. It is not just that the state seeks to produce a discernible realm of institutions and individuals, but that these institutions should behave in particular ways. We see this most clearly in legal processes of NGO registration (see Bolton and Jeffrey, 2008), but it is similarly conveyed through more subtle processes of NGO funding and regulation. These processes perform a dual manoeuvre: cultivating a sense of economic individualism through competitive funding processes while attempting simultaneously to present these organizations as indications of new forms of collaboration and inclusivity. Through the practices of NGOs, dispersed through meeting rooms, cafés, streets and offices, we see the productive effects of performances of state power. The behaviour of many institutions was shaped by a desire to conform to state-led concepts of civil society, while others sought to reflect alternative ideas of state sovereignty.

The politics of 'building civil society' was particularly acute in Brčko District. As discussed in Chapter Four, following the announcement of the Final Award the Supervisor suspended District elections until October 2004. In light of the nationalist outcomes in polls across BiH, his intention was to wait until political parties emerged 'with Brčko-based agendas'.[1] The Supervisor explained that this did not curtail democracy, as the proceedings of the institutions of governance (the District Government and Assembly) were open and transparent, allowing local civil society groups to witness deliberations and lobby councillors.[2] In the absence of elections, this demo-cratic model of transparency was dependent upon an audience: a sense of a viewer or viewers that could register and intervene in the political process in Brčko. Immediately we see in this language a separation, as the state is conceptualized as the formal processes of government and civil society as a compliant audience. Consequently, a series of initiatives were established to strengthen civil society in the District, such as the $3.1 million UNDP 'Brčko Local Action Programme' and a youth NGO programme by the German development agency *Gesellschaft für Technische Zusammenarbeit* (GTZ). In addition to these schemes a number of other donors funded NGOs unilaterally, often attracted by Brčko's unique status in BiH as a Special District. These included large-scale donors such as the United States Agency for International Development (USAID), the United States Department for Agriculture (USDA) and the Swedish International Development Corporation Agency (SIDA), in addition to smaller support agencies such as the Diana Princess of Wales Memorial Trust Fund.

One of the key empirical questions when studying NGOs is the fraught issue of what constitutes the existence of an NGO. It is for this reason that there is uncertainty about how many NGOs exist in Brčko District. When I first started fieldwork in BiH I drew on the work of Mercer (2002) to estab-lish four selection criteria to survey the number of NGOs: legal registration, and social welfare-orientated, non-profit-making and non-political (in the sense of non-party-political) objectives. Following the criteria mentioned above, eighteen NGOs were surveyed in Brčko. These organizations were often divided into a series of typologies by donor or regulatory agencies: funded/ unfunded, religious/secular, international/local or nationalist/multi-ethnic. But regardless of their numerous differences, they were all engaged in a struggle to attain funding from donors and recognition from those organizations in authority. It is these struggles that are explored in greater detail through a detailed examination of the accumulation of social and cultural capital.

5.2   Social Capital: The Autonomy of Civil Society

Robert Putnam (1993), drawing on Coleman (1968), defines 'social capital' as the accumulated value of groups or individuals working together. This has proved an enduring definition, particularly within neoliberal devel-opment circles. Fukuyama (1995), for example, collapses the concept of 'social capital' into a single notion of 'trust', convertible into 'economic cap-ital' in the form of increased profit. In 1996, the World Bank established the

Social Capital Initiative (SCI) to fund research exploring social capital as a development tool. The SCI research and associated World Bank programmes have come under sustained criticism, among other things for de-politicizing development and reproducing inequality (see Fine, 2002; Harriss, 2002). In addition, this work has also criticized the Bank for failing to incorporate Pierre Bourdieu's more socio-political understanding of social capital (Bebbington, 2004). In contrast to Putnam's communal trust and cooperation, Bourdieu defines social capital in terms of 'social connections' (Calhoun, 1993: 70) or 'group membership' (Bourdieu, 1987: 4). Rather than 'societal glue' (World Bank, 1996), this definition focuses on the ability of individuals and groups to accumulate capital.

Reflecting Bourdieu's interpretation of the term, an examination of the Brčko NGOs illustrates the struggle each organization undertook to accumulate social connections and convert these into economic capital (funding) and symbolic capital (legitimacy). While this shaped the relationships between NGOs and donors (discussed in detail below), one of the most profound effects of this struggle to accumulate capital was on the relationship between the individual NGOs. Rather than Fukuyama's (1995) 'trust', the struggle to accumulate social capital seemed to be spreading institutional insecurity. Those organizations with little funding suggested that the internationally financed organizations were merely 'keeping an office alive'[3] while, in turn, those organizations with long-term funding would accuse the smaller NGOs of 'existing only on paper'[4] or being allied to 'nationalist politics'.[5] A Firefly Youth Project representative explained that this insecurity was due to competition for funding, suggesting, 'if you talk to them about collaborating with each other to get double the funding instead of competing with each other, people become really suspicious'.[6] The issue of future funding was regularly greeted with a hushed 'no comment' by informants; information was seemingly embargoed until the outcome was known or perhaps until the closing date for proposal applications had passed. One NGO representative demonstrated this practice of confidentiality when explaining that they did not want to discuss the proposal they were submitting to the UNHCR as this could 'jeopardize'[7] their chances of success.

In addition to retaining institutional 'secrets', certain NGOs viewed other organizations as 'too political' to warrant collaboration. Particular scepticism was shown towards the 'Youth Coordination Body' (YCB), a contentious initiative of the OSCE and part of their democratization remit, designed to improve communication between youth NGOs and political parties in Brčko. On the one hand, political parties were broadly positive about the initiative, with the SDS representative suggesting it would be 'useful in the future as an institution of democracy',[8] while the *Socijaldemokratska partija* (Social Democratic Party or SDP) representative considered it 'nice, we understand each other, we support each other'.[9] On the other hand, many NGOs seemed more sceptical about the combination of political parties and NGOs. Some organizations had been part of the YCB at its inception, but had since stopped attending the meetings. A representative from 'Baza' Anti-terrorist Organization said that she had lost faith in the YCB after the Serb Radical Party had joined and it had 'turned nationalist'.[10] This wariness was shared by certain international donor organizations. A representative from the German development agency GTZ felt that the YCB was 'a problem because it involves political parties'.[11] These polarized responses suggest that membership of the YCB had costs and benefits for NGOs and political parties in terms of social capital. For the political parties, membership allowed them to work with NGOs, raising their profile both with their electorate and with organizations such as the OSCE, OHR and Western government donors as they were seen to associate with internationally recognized civil society actors. For the NGOs, however, the YCB proved a more ambiguous resource. For some, being tainted as 'political' could have repercussions in terms of gaining future funding, as the comments from GTZ attest, though for others the social capital of meeting with political parties could potentially lead to financial support from either the parties themselves or their nationalist sponsors.

The value of nationalist affiliation in social capital terms was evident in a number of the NGOs in Brčko. Both the Serbian Youth Association and the St Sava's Youth Association had gained funding from the RS Ministry for Sport and Youth. In the case of St Sava's Youth Association this funding was received because they 'had friends in the Ministry' that wanted to 'help our people',[12] while the Serbian Youth Association had gained funding for the 'process of registering in Brčko' their 'radio programmes' and their 'visits to Serbian Monasteries'.[13] The problem with this ethno-national social capital was not in the conversion into economic capital; through links with the RS Ministry for Sport and Youth repeat funding was mentioned by both organizations as a formality. Instead, the difficulty was converting this into other forms of social capital, such as contacts with international donor agencies or the OHR. Adopting the prefix 'Serbian', the Serbian Youth Association admitted, meant that 'certain foreign organisations don't want to deal with us or give us funds'.[14] This nationalist social capital appears mutually exclusive; drawing on the lucrative nationalist contacts ruled out funding from international sources committed to strategies of multi-ethnicity in post-conflict BiH. This also demonstrates the ways in which those organizations holding symbolic capital (in this case, the OHR) could influence the exchange values of different forms of capital, specifi-cally by placing a premium on 'multi-ethnicity'.

It is, however, important to note that for both 'international' and 'nationalist' social capital, the exchange values, bound as they are to the discursive priorities of those in authority, are not fixed in time and have shifted with political expedience and geopolitical circumstances. As an example, a representative of the SDS, who had previously worked in the RS Ministry for Sport and Youth, discussed with some pride the fact that his political party had recently held a seminar funded by the US Embassy, the first time such cooperation had occurred.[15] This kind of collaboration would have been unthinkable in the past, particularly considering the mutual animosity between the US government and the Radovan Karadžić-founded SDS. It seems that the discourse of multi-ethnicity has, on a BiH-wide scale, diminished in importance over the perceived move from post-conflict reconstruction to the *realpolitik* of economic transformation and international withdrawal. This shift, or narrowing, of priorities was demonstrated in the interview with the Brčko Supervisor. When questioned about the issue of the RS directly funding NGOs in Brčko, he responded: 'It is not in my mandate to worry about that, as long as they are not breaking Brčko laws. You know, I think the NGOs should be free to scrounge money wherever they can.' The concerns of the Supervisor, as enshrined in the Final Award, lay rather in the 'government financial area – there has been a failure by the RS to pay its obligations to the District and they tried to privatize some companies in the District without consulting us first'.[16] These comments are suggestive of the valorization of economic discourses by the Supervisor and the Final Award, marking a shift from 'post-conflict' concerns of inter-ethnic reconciliation. In this way, the 'negative' social capital of nationalist discourses has lessened as processes of neoliberal economic governance have taken precedence over concerns of reconciliation.

This account of the effects of funding and regulator regimes on the behaviour of NGOs challenges official discourses of 'building civil society'. Presenting the social connections of Brčko NGOs in terms of social capital, brings to light the shifting value placed on different relationships. In contrast to the conception of 'social capital' held by Putnam (1993) and the World Bank (1996), the struggle to gain funding in Brčko has encouraged competition and suspicion between the NGOs. While civil society is defined within certain policy scripts as an 'autonomous' group of organizations, the autonomy in this case is not from agencies of authority but conversely between each individual NGO. We see the emergence of a series of 'vertical' relationships between NGOs and donor organizations, while 'horizontal' connections to other civil society agencies are viewed with suspicion. The process through which NGOs are bound into closer ties to donors and the agencies of the state is illuminated in detail through a discussion of the accumulation of cultural capital.

5.3   Cultural Capital: 'Don't Just Ask for Another Copy Machine!'

Just as the social contacts of the NGOs have been mediated through the exchange of *social* capital, the embodied traits and behaviour of NGOs have been shaped through the accumulation and circulation of *cultural* capital.

Bourdieu (1986: 248) describes cultural capital as 'the aggregate of the actual or potential resources which are linked to possession of a durable network of more or less institutionalized relationships of mutual acquaintance and recognition', though he also offered the more concise definition of 'informational capital' (Bourdieu, 1987: 4). Consequently, it has been used as a theoretical tool to explain the role of education in reproducing social divisions (Painter, 2000: 240). Harker (1990: 87) suggests that as economic institutions are structured to favour those who already possess economic capital, 'so our educational institutions are structured to favour those who already possess cultural capital'. Similar processes of social reproduction were identified within the NGO sector in Brčko District, as funding regimes 'rewarded' organizations with particular cultural capital at the expense of others. The interlinked traits defining the cultural capital of NGOs are examined below in three sections: ability with the English language, knowledge of donor discourses, and skills to negotiate the bureaucratic demands of proposal writing.

A number of NGO informants lamented the importance of English to gaining donor funding. A representative from Firefly Youth Project suggested that she had 'never met a donor who took any funding proposals in Bosnian. Even if they have their offices here – a local person won't decide about donations so their boss must read it and I don't think he will translate it for us.'[17] The proposal-writing process, however, did not simply demand knowledge of English but expertise in the particular technical style demanded by funding agencies and government donors. Regardless of the level of skill of spoken English, the rigorous demands of proposal writing led to Bosnian NGO members being categorized as 'non-native English speakers'. A US expatriate USAID representative in Sarajevo explained her 'frustration' at poorly written English:

> It is inappropriate to use the active voice here in Bosnia, OK, so everything is 'the youth policy group was formed', by who? OK. 'A meeting was held where something was decided', well what was decided? So those of us who speak English as a first language and have been taught that you don't write or think or speak in the passive unless you intentionally want to soften something [...] are very frustrated. (Representative of USAID, Sarajevo, 29 May 2003)

A US volunteer to Food Security in Brčko confirmed this native-speaker/local binary, stating that the headquarters in Washington had become 'very concerned'[18] with the quality of the reports written by Brčko staff. On his arrival in Brčko his core duties involved preparing official reports and writing proposals. This task allocation is ironic, given that the head office in Washington had recruited three of the four Bosnian members of staff on account of their skills at English. Indeed, two of the appointments had been given to the former translators of expatriate staff, thereby demonstrating the premium placed on English language over technical knowledge of the NGO's activities, in this case agricultural management. Thus, the staff were considered to hold a proficient level of English for regular communication with the head office in Washington and for OHR coordination meetings, but not adequate for the 'official' channels of funding proposals and project reports. This suggests that the level of language skill requisite for donor funding appears to be unattainable to Bosnian staff members. While there were distinct advantages in terms of cultural capital of an individual speaking English, this binary ensured that there was still greater capital in an NGO retaining links with, or employing, 'international' or 'non-local' staff. Where discourses of civil society circulating within organizations such as the OHR focus on the importance of grassroots mobilization (see above), this evidence is suggestive of the value placed on retaining 'native English speakers' to assist with funding applications and administration.

These benefits had not escaped the attention of individual donors as they attempted to enact policies to counter this binary. In particular, international donors had begun to value local NGOs over international organizations, in a bid to assist only those organizations that would be sustainable in the long term. This approach demonstrates the symbolic capital possessed by donor organizations as they invest 'being local' as a trait with particular cultural capital. Through the process of 'localization' donors sought to develop 'sustainable locally run, voluntary organizations',[19] a policy explained to me, perhaps ironically, by an Italian employee of UNDP in Sarajevo. This policy discourse was reflected in the UNDP Brčko Local Action Programme, where preference was given to funding local organizations, as part of a broader aim to 'capacity-build the Brčko NGOs'.[20]

In light of this donor priority, many of the NGOs in Brčko exhibited a compromise between 'localization' and the donor requirement for competent English speakers. Through such a compromise, hybrid NGOs emerged which were both 'international' and 'local', and they deployed these identities in order to maximize their capital. For example, Firefly Youth Project received three-year funding from the Diana Princess of Wales Memorial Trust Fund directed towards creating a sustainable local NGO. At the completion of this funding cycle (in 2002) the organization changed its name to *Omladinski projekt 'svitac'* (a direct translation of Firefly Youth Project) and began the process of registering in Brčko. Despite this new local organization, Firefly Youth Project still exists in the UK and continued to draft money to *Svitac* in order to meet its running costs, while also supporting the organization through assistance with proposal writing. The international contacts retained by *Svitac* also enable continuing visits by UK volunteers. These were ostensibly to implement short-term projects (such as music lessons or art projects), though the programme manager acknowledged an additional benefit that 'they can go to all the OHR meetings for us'.[21] It seems that in the case of Firefly/*Svitac* this was simulated self-sustainability, where 'localization' was mobilized to attract donor funding, though for logistical purposes international links were retained. This example demonstrates the way in which labels such as 'international' and 'local' are reformulated and deployed by NGOs in order to maximize their cultural capital.

This experience of 'localization' reflects how the funding system required more than simply knowledge of the English language, but also an awareness of the priorities and discourses of development agencies. As Cornwall and Brock (2005) note, development orthodoxies are communicated within a shifting discourse of buzzwords, and proficiency with this vocabulary is vital for retaining donor funding. Many of the NGOs interviewed in Brčko felt that they could not 'keep up' with changing donor agendas and found the shifting priorities 'bewildering'.[22] A representative of Firefly/*Svitac* explained that when it came to writing proposals they 'didn't know what donors wanted to hear – what we really need is one [international] volunteer who will sit with us and write one good proposal'.[23] This notion of 'one good proposal' suggests that cultural capital can be held in material form, a pro-forma document that could gain repeat funding from multiple donors. This also pointed to the way in which NGOs in Brčko often divorced needs assessment from the technical process of writing a donor proposal. Contact with donors was framed in terms 'they wanted to hear' as opposed to reflecting the ideologies or activities of the NGOs. As with language skills, the importance of being able to understand and reproduce donor discourses influenced the recruitment policies of the NGOs, since individuals with experience of international organizations were perceived as preferable to those with specialist technical knowledge. As a result, individual cultural capital was not accumulated through formal education or qualifications, but rather through knowledge of English and experience of working with NGOs and international organizations within Brčko.

In addition to the requirement for language skills and knowledge of donor discourses, a final aspect of cultural capital that shaped the NGO sector was the ability to negotiate the more mundane bureaucratic requirements of the proposal-writing process. Many of the NGOs did not have designated office space, for example the Bosniak Women's Association met in a nursery school in Brodusa and the Grčica Youth Association met in a roadside café on the outskirts of Brčko. Concerns over office space led to a feedback loop developing, where office equipment (necessary to negotiate the bureaucratic demands of donor funding) became a key feature of NGO funding applications in Brčko. This cycle was reflected in the exasperated comments of the Head of OSCE Brčko when he said, 'I am always telling NGOs "don't just ask for another copy machine!"'[24] This over-bureaucratization suggest that offices themselves hold cultural capital, becoming a key feature of 'being an NGO'.

A second facet of the bureaucratic funding system relates to the length of time it took to receive confirmation that a project could be financed. This was particularly evident in the case of the agricultural NGO Food Security. Though Food Security benefitted from long-term funding from the USDA, each of its projects had to be approved by their head office and a second partner NGO based in a field office in the Bosnian town of Tuzla. In addition to the language difficulties discussed above, gaining approval for a project involved a bureaucratic process that could last over six months. According to one Food Security official, this had the consequence that projects became 'more conservative' as no one wanted to 'end up wasting time on something that is going to end up not getting approved'. This official added, 'you couldn't ask [the head office] a question as if it dropped off the bottom of her Outlook screen she would never answer it'.[25] In order to ensure funding, and to expedite the bureaucratic process, the projects of Food Security became smaller in scale and repetitive.

These examples demonstrate that the donor system has valued particular cultural traits that have gone on to shape the attributes, activities and politics of the NGOs in Brčko. In the case of English language skills, this has continued the trend observed with social capital in binding NGOs ever closer to international organizations, leading to compromised 'hybrid localization' and the prioritization of fulfilling bureaucratic demands. Such practices serve to restrict the transformative qualities of NGOs, heralded as an attribute of civil society (not least within OHR documentation; see OHR, 2000). But significantly, we see in the practices and decisions of these organizations considerable space for agency: they are not duped by agencies of donors but rather see adhering to their demands as a form of compromise that ensures institutional survival. This finding reflects Tania Murray Li's (1999, 2007) conclusions, drawn from her ethnography of development interventions: there is no singular outcome to the governmental interventions in Brčko, but rather we see a range of situated practices that seek to both inhabit and challenge dominant ideas of the state.

Supporting the arguments of scholars such as Mawdsley *et al.* (2002) and Dolhinow (2005), the funding system in Brčko has privileged individualistic and conservative organizations which seek to reproduce the status quo rather than acting as loci of dissent or radicalism. This has a series of implications for how we conceptualize the distinction between 'the state' and 'civil society'. The funding regimes of donor organizations should be understood as improvisations of the state, as a series of institutions seek to cultivate a set of compliant organizations carrying out projects that strengthen the idea of the Dayton Bosnian state. Quite explicitly, the practices that are enshrined for international funding are those that strengthen the idea of a coherent and integrated Bosnian polity, through projects directed towards reconciliation, multi-ethnic education and economic development. As Navaro-Yashin (2002: 136) found in the case of Turkey, the line between civil society and the state becomes blurred to the point that an analytical distinction seems 'obsolete'.

5.4   Beyond Gentrified Civil Society: Roma and Mjesne Zajednice

We waited by the verge of the road to Tuzla in Broduša in the bitter cold. Sanela [Research Assistant] seemed unsure of where we were to meet the representative from the Roma Association, a group I had previously been told 'did not exist' and were 'impossible to get hold of'.[26] We had managed to call the President of the Association on a mobile phone and he had told us to come at a specified time and wait in the centre of Broduša. After twenty minutes we called again and had no response. We were beginning to give up hope when a large battered blue lorry rumbled to a halt next to us and a man opened the door and said 'Fahrudin sam, hajde' [I'm Fahrudin, let's go]. He was a giant of a man with hands like satellite dishes covered with tattoos. We drove through Broduša, past many lorries dropping off supplies for the permanent reconstruction work that is taking place in this part of town that was once the frontline of the conflict. After ten minutes the asphalt ended and the road turned into a quagmire. It was very misty as the lorry stopped where the minefields stretched out marking the end of town. We headed for a house which, like many others, had only its ground floor rebuilt. Fahrudin beckoned us in. We were ushered through a bare front room (apart from a very colourful carpet) and, after going through a curtain, up some concrete steps. We were in a living room upstairs, where there were no windows and it was extremely cold. Sitting in the room already were two other men, as it turned out the Vice-President and Treasurer of the Roma Association. They seemed resigned to the lack of help from the government, as they weren't getting any money. It struck me that they didn't want to 'play the game' like other organizations. They didn't talk in buzzwords, they just wanted to stress the need for better sanitation and living conditions for the Roma. In their discussion of the conditions, and from what I had seen on the journey in the lorry, nothing had moved forward, it seems that Dayton has overlooked them and they haven't been able to go through the usual channels. Halfway through the interview an old lady entered the room [Fahrudin's mother] and sat on a carpet-covered sofa next to the window. She kept looking across and smiling while gesturing as if to rub her arms saying 'zima, zima' [winter, winter]. She said she had twelve children and what they really needed was a disco, and then laughed doing a funny dance with her arms. It was a real contrast to the three men. They explained that they couldn't even gain support from the network of Roma Associations in Tuzla, Sarajevo and further afield because they are perceived as rich because they come from

Brčko District. But Brčko District Government isn't helping them out. So it seems that both the 'informal' and the 'formal' networks are failing them.

Extract from field journal, 27 February 2003

This account of a trip to the Roma Association in Broduša, on the outskirts of Brčko, is a reminder that the construction of 'civil society' is simultaneously a performance of the state. The absence of Roma groups from the gentrified civil society discussed above illustrates how the creation of the Dayton state has relied upon particular assumptions of Bosnian society. Most significantly, Roma groups are not represented within the ethnic equations and weightings of official government positions as set out in both the Dayton Agreement and the Final Award in Brčko. Indeed, the presence of Roma is doubly threatening: they challenge the 'fault lines' of the GFAP constitution while failing to behave in a way that corresponds to the established conventions of gentrified civil society. The perceived threat of the Roma is, of course, not restricted to BiH, as indicated by the recent evictions of the Roma groups in France (see Radu, 2011). At the heart of this concern is that the Roma challenge the very performances of stability that characterize improvisations of the state. As indicated in the experience above, Roma associations did not send representatives to OHR coordination meetings and (unlike many other organizations) they took no part in the negotiations to reduce the NGO registration fee (see Bolton and Jeffrey, 2008). The language of the meeting with the Roma Association differed markedly from other interviews; there was no mention of applying for donor funding, of speaking English, of printing business cards or establishing an office. The organization did not appear to be in a struggle with other organizations over such objects of social or cultural capital, and, this being so, it did not seem interested in displaying the traits that constituted 'being an NGO' in Brčko. During the interview many of the connections they found of greatest value were transnational links to Roma groups outside BiH, and some outside the states of the former Yugoslavia.

The Roma Association was not the only example of such ungentrified civil society in Brčko District. Another group of organizations that were acting as focal point for community participation were the MZs. As discussed above, these territorially constituted organizations emerged from the 1974 Yugoslav Constitution; they were the lowest strata of local government and conceived as a form of direct popular association (Pusić, 1975: 137). They were small-scale community organizations, with eighty-two covering Brčko District (Rosenbaum, 2001). Though the 1974 Constitution did not grant MZs revenue-raising powers, in practice this became necessary as the Yugoslav Federation began to weaken. Consequently, each Yugoslav municipality became a federal arrangement of MZs (see Alcock, 2000: 92). The role of MZs in Brčko changed after the conflict. Following the Final Award in Brčko District they were formally disbanded as, according to an OHR

Political Officer, 'the authority of Brčko District Government cannot be sub-divided, and so the *mjesne zajednice* cannot have any governmental function'.[27] This differs from the RS and the Federation, where the MZs have retained a government role, collecting taxes and indicating areas of community need (Jones, forthcoming). Since their formal dissolution their role in Brčko had been more difficult to define, as they continued to function informally as loci of community organization, particularly in the field of refugee returns (see Dahlman and Ó Tuathail, 2005a). Though MZs were traditionally territorially defined, following the displacement of Brčko's population during the conflict whole MZs relocated to function as 'MZs in exile'. In an interview, a Bosniak displaced person (DP) explained the way in which the MZ from Klanac relocated to the village of Rahić following the violent displacement of the urban population to rural areas of the District:

DP: These old MZs never stopped working, for example MZ Klanac leadership was the same but in Rahić [a village in the south of Brčko District]. They had their office there, they had people who would come and join and have fun, well not fun, but have meetings or whatever. They organized events.

AJ: So people displaced from Klanac all over the District would go to the office in Rahić?

DP: Yes, and they organized reconstruction. The Serbs in Klanac in town formed their own MZ and have stayed.

(Excerpt from an interview with a Bosniak DP,
Brčko, 22 April 2003)

As this exchange suggests, the MZs became conduits for information between state agencies (such as the OHR and the Brčko District Government) and returnee groups. While NGOs were apparently misunderstood and perceived as 'international', MZs were reportedly 'much easier to understand [...] because it was like that before the war and people can see MZs are fighting for them. [...] [T]here are people every day going to MZs with requests for something or another'.[28] It seems that the continuity of MZs, from pre- to post-conflict, granted them a variety of symbolic capital, manifest as legitimacy and trust, which the gentrified NGOs could not accumulate, regardless of funding, capacity building or institutional support. Often NGOs drew on this symbolic capital to advance their work; the Italian agricultural NGO *Cric*, which was based in a rural part of the District, described MZs as 'vital partners'[29] when working with rural communities.

Despite these successes, there were many concerns among the international agencies about the political role played by MZs, as alluded to by a former OHR official:

The most important role of MZs since the war has been the dissemination of information. It is about dissemination of information both upwards and down. And the problem is that since the MZs themselves will end up themselves being politically aligned, the information that is being passed will always go through a filter at the MZ leadership stage. (Interview with former OHR official, Brčko, 2 June 2003)

While certain MZs in Brčko District were described as too political to be regarded as 'civil society' actors, others were embraced as partners within community development projects. The problem that was often communicated in the interviews was the difficulty of knowing which of these kinds of MZ you could be dealing with, as explained by an OHR Political Officer:

[...] they are a bit of a Jekyll and Hyde. [...] They [MZs] really are the most committed form at present, taking care of the residents within that area. They have, in that sense, very legitimate concerns. But there are examples, such as Klanac, which are totally dominated by political parties, by party politics. I think the problem lies, in fact, when it comes to [pause] coming from a position where they had very strong links with the government when they had all these competences at the local level and they had an understanding with the government on how things are done and what to expect. The same kind of scenario has embedded itself, as it does in most relationships, that the politics take over and becomes an issue of you having the power and you putting your man into government. And this is something that we wanted to break. (Interview with an OHR Political Officer, Brčko, 24 May 2003)

These perceived dangers of MZs as potentially 'Hyde'-like political organizations were also reflected in the language of the USAID-funded District Management Team report on municipal assistance for Brčko District compiled in 2001:

By not fully appreciating the potential obstacle that the MZs could have been, and still may be, to the process of governance consolidation, a false sense of security regarding the stability of the institutions that have been established has perhaps been created. In this regard, efforts should be initiated to create ethnically integrated, community-based units of the District Government which can serve as a part of the Brčko governance structure. In all probability, at least some of the leadership of the MZs can and need to be integrated into such an effort. One alternative that might be initiated by the District Government is the establishment of 10 to 15 geographically based (if necessary, they could be gerrymandered in such a manner as to insure ethnic diversity) elected or appointed neighborhood councils. Such councils, while remaining subordinate to the District Government, could be delegated authority to allocate modest financial resources to facilitate community improvements and possibly given some degree of authority in the areas of planning and zoning. (Rosenbaum, 2001)

These candid remarks point to the level of international engineering which was considered necessary to depoliticize and regulate the activities of MZs. This strategy was evident in the UNDP BLAP's attempt to create 'Community Development Groups' as rival non-political local community associations, in a similar vein to the 'neighbourhood councils' suggested by Rosenbaum (2001) above. The Programme Manager of UNDP BLAP explained enthusiastically that 'the MZs were not representative, so we approached the communities and told them about representation'.[30] According to a UNDP Project Officer, the process of 'improving representation' involved ensuring that the Community Development Groups included 'representatives of youth, farming associations and women'.[31] There have, however, been doubts in certain quarters as to the viability of 'creating' new systems of communal organization, as articulated by a former OHR Official:

> The MZs are much more experienced with dealing with the international community than the international community is with dealing with MZs. So, they will, I have been here a long time and they will quite openly, openly as far as I am concerned, they will recruit members of the other ethnicity in order to make them more attractive to donors and NGOs, they know that absolutely. One of the MZs I deal with is 100% Serb, but they did have one mixed marriage and I think there was a Muslim that lived on the border of the MZ. And the MZ leader was hilarious in including their name so that they got [a] Muslim, a Croat and a Serb! The cynical manipulation of the international community is quite amusing to watch. (Interview with former OHR official, Brčko, 2 June 2003)

This MZ-led representation was reflected in the formation of the UNDP 'Community Development Groups' in eleven localities in Brčko District. The evaluation report of UNDP BLAP described the selection and setting up of these development groups as 'particularly challenging for a number of reasons including local mistrust towards outsiders and a sense of dependency and entitlement due to previous humanitarian assistance for which nothing was asked in return' (UNDP, 2001: 23). In light of these potential difficulties I asked the UNDP Programme Manager how they had approached the target communities and she laughed and exclaimed, 'Oh, mainly through the MZs!'[32] In supporting the sentiment of the former OHR official, this comment reveals a sense of pragmatism which permeated other areas of the UNDP Programme. In the glossy Community Development Strategies produced for each of the eleven localities the quantitative data is divided in terms of MZs and the groups define themselves as 'Local Community Development Groups', suggesting almost a subset 'development group' of the *mjesne zajednice*. It seems that although the UNDP went to some lengths to avoid the discourse of MZs, tainted as they were with uncertainty and political associations, they remained a key tool for organizing community participation.

The process of rendering the MZs legible as Community Development Groups appeared to be more for the benefit of agencies of authority than for local people. In a move related to the classification of MZs, in 2004 the OHR drew up a Law on Local Communities. Like the Law of Associations, the intended outcome of this legislation was the legal registration of all the MZs, coupled with a formal rejection of their governmental function. An OHR Political Officer described this process:

> The Law on Local Communities establishes a reality that government competences have been taken away from local communities and so now in order to reassert their power they need to depend on something other than simply institutional cooperation with the government. They need to build up as grassroots pressure which is the only way of holding the government accountable. (Interview with an OHR Democratization Officer, Brčko, 24 May 2003)

This comment portrays how the new Law of Local Communities was being designed to place institutional distance between the government and the MZs. This rhetoric is reminiscent of the discourses of institutional autonomy, rendering civil society as distinct from the state. The familiarity between MZs and the District Government was, according to the OHR, a barrier to the celebrated democratic attribute of 'transparency':

> The way that authorities view MZs and include them is because the relationship is much more defined with the MZs than it is with the NGOs. Their patience with them is just a testimony that they are aware what is going on, who this MZ belongs to politically and their political infighting. And if these political bodies exercise influence on the government over the allocation of resources then we are in a very non-transparent system. (Comment made by OHR Democratization Officer, Brčko, 24 April 2003)

This comment demonstrates the broad desire among the agencies of the state to convert MZs to NGOs so as to render the organizations 'more transparent', just as the District Supervisor had seen the role of NGOs as viewing the transparent democratic performance of the District Government (see Chapter Four). The UNDP Programme Manager explained that they would have a small fund that they could apply to for the funding of projects.[33] The Proni Regional Coordinator described a programme they were organizing to train MZ leaders in drawing up logical frameworks for funding proposals.[34] Such strategies coalesce to represent a broad tactic of gentrifying the MZs into NGOs, to expose them to the same struggles for capital and attempt to render them 'autonomous' from government intervention. Such an objective operates with a narrow interpretation of governmental action, as the evidence from this chapter suggests that the gentrification of MZs into NGOs would not lead to the autonomy of such organizations but rather qualitatively alter their enrolment into the operation of state power. In contrast to the MZs as recognized governmental agents, the production of 'NGOs' allows the illusion of autonomy, to reproduce the notion of a separate state and civil society.

The examples of the Roma Association and the MZs present challenges to the discourse of state–civil society relations set out by intervening agencies in Brčko. The experience of these organizations illuminates the incomplete and plural nature of state improvisation, as both the Roma and MZs cultivate alternative spatialities of political authority. In the case of the Roma, they are establishing both local and transnational links to secure necessary funding and welfare. For these groups, the rigid ethnic matrix set down at Dayton has fostered exclusion from the political organizations of government. Since the District Supervisor conceived of civil society as an 'audience' for the formal deliberations of the local government, the lack of representation of Roma groups renders this function obsolete. In the case of MZs, they are constitutional legacies from Yugoslavia, and hence cultivate forms of local (ward-level) affiliation, often structured around ethno-national identity. The ambiguity within international agencies as to how to categorize MZs is another example of the paradox of the political geography of Dayton. As discussed above, the MZs are often constituted of a single ethno-national group, a legacy of either the nature of Yugoslav rural demographics or the homogenization of urban districts during the conflict. While the partition of Dayton into the RS and the Federation supported such ethno-national distinctions, this works in tension with the simultaneous desire to cultivate a multi-ethnic citizenry. In the case of MZs we see the institutional expression of this tension: they are respected and trusted institutions within Bosnian society, but their vision of political community does not cohere with the improvisation of the Dayton state. It is to bridge this gap that we see the struggle by the UNDP and the OHR to modify MZs, to render them legible to the instruments of funding and regulation established first at Dayton and further by the Final Award in Brčko.

5.5 Conclusion

This chapter has examined how the imagined distinction between state and civil society is blurred and ambiguous in practice. Using the example of Brčko District, the account has demonstrated the environment in which NGOs operate, a context that sets an exchange value for particular social and cultural traits. The evidence from Brčko is suggestive of the emergence of particular professionalized NGOs, where financial survival is predicated on defending institutional boundaries and replicating donor priorities.

These practices have had profound political implications for the position of NGOs. Rather than acting as spaces of alternative political action or communal participation, as set out in ideals of civil society, these organizations have forged close links either to international organizations or to nationalist political parties in attempts to secure funding and legitimacy. This may seem a technical or instrumental problem, but it has profound implications for understandings of politics. The equation of civil society with a very narrow form of liberal democracy (with an emphasis on rationality and individualism) invalidates alternative forms of political life, suggested here through the actions of MZs and Roma associations. As Chantal Mouffe (2005) has convincingly argued, this approach to democratic governance serves to close down political debate, as meaningful alternatives are not tolerated within the democratic model. In the case of Brčko, the gentrification of civil society seeks to create rational and individualistic political actors that are compliant with the prescriptions of the improvised state.

These conclusions have tangible implications for the development of 'political parties with Brčko-based agendas', a stated aim of the OHR in Brčko District and the reason given for the suspension of local elections. Rather than opening dialogues with the local population, the NGOs surveyed in this study appeared to be striving to forge links with agencies of the state. These practices seemed to have had the consequence of further removing NGOs from the communities they serve, reproducing the discursive priorities of intervening organizations and nationalist political parties. This argument echoes the conclusions of Ferguson (2006: 96), that while failing to produce 'Brčko-based agendas' the power of the Bosnian state is expanded and entrenched. In addition to the continued use of 'Bonn Powers' by both the District Supervisor and the High Representative to remove government officials and pass laws in BiH, the detailed qualitative evidence in this chapter demonstrates the mechanisms through which international agencies have continued to exercise symbolic authority over associative life.

Notes

1 OHR Brčko Democratization Officer, 13 September 2004.
2 Brčko Supervisor, 24 April 2003.
3 Representative from Grčica Youth Association, 22 November 2002.
4 Representative from Firefly Youth Project, 15 September 2002.
5 Representative from 'Baza' Anti-terrorist Association, 22 November 2002.
6 Representative from Firefly Youth Project, 15 September 2002.
7 Representative from Centre for Legal Information and Help, 22 October 2002.
8 Representative from SDS, 14 April 2003.
9 Representative from SDP, 16 April 2003.
10 Representative from 'Baza', 22 November 2002.
11 Representative from GTZ, 4 June 2003.

12   Representative from St Sava's Youth Association, 3 December 2002.
13   Representative from Serbian Youth Association, 11 September 2002.
14   Survey of the Serbian Youth Association, 21 October 2002.
15   Representative of the SDS, 14 April 2003.
16   Brčko Supervisor, 24 April 2003.
17   Representative from Firefly Youth Project, 27 March 2003.
18   Food Security Official 'A', 18 September 2004.
19   UNDP official, 29 May 2003.
20   Projects Officer at UNDP, 15 April 2003.
21   Representative from Firefly Youth Project, 27 March 2003.
22   Representatives from Bosniak Women's Association, 21 February 2003.
23   Representative from Firefly Youth Project, 27 March 2003.
24   Head of Brčko OSCE, 17 October 2002.
25   Food Security Official 'A', 18 September 2004.
26   These comments had been made in an interview with a representative from Firefly Youth Project, 15 September 2002.
27   Discussion with the OHR Political Officer, Brčko, 13 September 2004.
28   Interview with a Bosniak DP, Brčko, 22 April 2003.
29   Interview with a representative from Agricultural NGO Cric, Donja Skakava, 18 March 2003.
30   Interview with UNDP BLAP Programme Manager, 18 February 2003.
31   Interview with UNDP Project Officer, Brčko, 15 April 2003.
32   Interview with UNDP BLAP Programme Manager, 18 February 2003.
33   Discussed in an interview with the UNDP BLAP Programme Manager, Brčko, 18 February 2003.
34   Discussed in an interview with the Proni Regional Coordinator, Brčko, 7 May 2003.

Chapter Six

Enacting Justice

We are in the middle of nowhere with [war crimes trials]. Yes, some things are happening in The Hague, but it is time to exit this system. New institutions like the new War Crimes Chamber in Sarajevo are great. But they are often empty shells with no content.

Interview with Dragan Musić, Human Rights Centre,
Sarajevo, 7 October 2009

This chapter examines the politics of institution building at the Bosnian state level, through the lens of the establishment of the State Court of Bosnia and Herzegovina (CBiH) between 2000 and 2009. The comments from the representative from the Human Rights Centre reflect a form of despondence I had heard on a number of occasions during fieldwork in BiH in 2007 and 2009. Namely, that the judicial system in BiH was caught 'in the middle of nowhere' between the high-profile activities of the International Criminal Tribunal for the former Yugoslavia (ICTY) in The Hague and attempts to establish a single judicial system, including a war crimes chamber, in BiH. It seems that international celebrations of ICTY's role in holding war criminals accountable for their actions has not been shared by all members of the Bosnian public. Instead, the distance of the ICTY and its necessary selectivity in indicting war criminals has fostered a sense of both an absence of accountability for the crimes of the past and a lack of state judicial sovereignty. This was not simply a public perception; as late as 2003 Paddy Ashdown, then High Representative, told the UN Security Council that he was addressing BiH's 'lawless rule' (United Nations Security Council, 2003).

The Improvised State: Sovereignty, Performance and Agency in Dayton Bosnia, First Edition. Alex Jeffrey.
© 2013 John Wiley & Sons, Ltd. Published 2013 by John Wiley & Sons, Ltd.

This tension between the activities of the ICTY and the absence of a Bosnian state court reflects a more complex geopolitics that lies behind international judicial interventions. In line with the theoretical and empirical approach of improvisation developed over the previous chapters we need to challenge a simple relationship between 'international' and 'domestic' institutions. Specifically, the recent establishment of a state court does not necessarily equate to the 'localization' of the judicial process. The creation of the CBiH in 2005 has required intense international supervision, from the imposition of the Law on the State Court of Bosnia and Herzegovina by the OHR in 2000 through to recruitment of an international judiciary to work alongside Bosnian judges. But the resulting institution is not merely a version of the ICTY transposed to Sarajevo. While the ICTY is an *ad hoc* tribunal established to follow a narrow mandate, the CBiH is a state court directly accountable to the Bosnian citizenry. Consequently, in tandem with the process of creating the formal legal jurisdiction of the Court there have been attempts to establish its legitimacy with the Bosnian people. In the case of Brčko District we saw that the improvisation of the state was structured around attempts to influence individual perceptions: to make people 'think' in terms of Brčko District. Similarly, the creation of the Court has involved a series of initiatives directed towards shaping popular participation and understandings of judicial processes in BiH.

By examining the formation and practices of the CBiH this chapter will explore attempts to establish the legitimacy of this new judicial institution. In particular, I will focus on the mechanisms used to foster popular participation in, and support for, its processes. This approach draws the analysis back to the enrolment of 'civil society' into the operation of new state institutions. In this case intervening agencies established a Court Support Network (CSN) of NGOs to assist in the trial process and spread public understanding of the activities of the new court. Part of the official explanation for this process has been a desire to cultivate autonomous NGO activity that supports the formal court activities. While the practices of the CSN illuminate a set of compliant organizations that are dependent upon the resources and logistical support of the court, there is also evidence of alternative spaces of justice emerging. Through interviews with CSN members, and participation in some of their activities, the lived experience of 'localizing' international law illuminates a set of tensions between forms of retributive and trial-based justice, and more deliberative and restorative approaches.

The chapter is divided into four sections. The first section examines the contested geopolitics that frame the establishment of the CBiH. In particular it explores the barriers that have existed to establishing a single legal institution in BiH, from the geography of the Dayton state to the forms of transitional justice advocated during and after the conflict. The second section explores the political opposition to the establishment of the Court that has been mounted in BiH. This discussion focuses in particular on the contested histories of the court site itself, since these debates animate wider tensions concerning commemoration, spatial justice and victimhood in contemporary Bosnian politics. In the third section the discussion focuses on the mechanisms used by the CBiH to build the legitimacy of the court with the wider Bosnian public. These initiatives, collectively referred to as 'public outreach', have enrolled civil society organizations across BiH in order to engage the Bosnian public with the aims and activities of this new judicial institution. The final section examines the alternative spaces of justice that are being promoted through these activities. While the concept of 'public outreach' may infer a rather transactional relationship between state and society, the agency of the NGOs involved has seen the operation of a range of alternative approaches to transitional justice.

6.1   Spaces of Justice

[The] over-riding priority, as we have discovered in Bosnia, in Kosovo, in Afghanistan and now Iraq, [is] establishing the rule of law – and doing so as quickly as possible. Crime and corruption follow swiftly in the footsteps of war, like a deadly virus. And if the rule of law is not established very swiftly, it does not take long before criminality infects every corner of its host, siphoning off the funds for re-construction, obstructing the process of stabilisation and corrupting every attempt to create decent government and a healthy civil society.

Ashdown (2003: 8)

The protracted attempts to establish a state court in BiH provide a further illustration of the improvised nature of sovereignty claims after Dayton. In these comments Paddy Ashdown, then High Representative, identifies a single legal system as a catalyst for the creation of rule ('decent government'). Reflecting the findings of the previous chapter, Ashdown also cites the significance of an internationally sponsored civil society to attempt to build a democratic state system. The creation of a single rule of law in BiH had been a priority of the OHR over the preceding decade; Annex IV of the GFAP had required the establishment of a single judiciary in BiH for the purpose of reviewing state-level administrative decisions and resolving constitutional disputes (see Council of Europe, 2001). But attempts to create a state court had met with considerable obstacles. First, as we have seen in the previous chapter, the line between legality and illegality has often been blurred in BiH as the formal institutions of the state have been infiltrated by the practices and personnel of organized crime (see Andreas, 2008). As we will see below, a considerable resistance to the creation of a state court in BiH has come from the current President of the RS, Milorad Dodik, who is opposed to inquiries into his own role in embezzlement and fraud. But the more profound political barrier to the creation of a single judicial entity has been the geography of the Dayton BiH. Annex IV of the GFAP required the creation of a state court and a prosecutor's office in BiH, though, as we have seen, this demand was set alongside the partition of BiH into two Entities (Federation and RS) and, ultimately, Brčko District. As a consequence, three distinct legal systems emerged across these three jurisdictions. Since legal territoriality underpins conceptions of modern state sovereignty (see Delaney, 2001), this fragmentation is itself a profound challenge to the unity and coherence of the Bosnian state.

But endemic corruption and the partition of the Bosnian state feed into a third obstacle to the creation of a state-level court. As discussed in previous chapters, the implementation of the GFAP has involved international supervision of executive and legislative processes across BiH. So-called 'Bonn Powers' (see Chapter Three) have been used by successive High Representatives, though to differing extents, to pass legislation or dismiss elected officials with the stated objective of implementing Dayton. As Caplan (2005) notes, the use of privileges has created an 'accountability gap' where international officials have the authority to make executive and legislative decisions without answering for their actions to electoral constituents (see Chapter Four). Within this framework, law becomes a tactic of the OHR, through which political objectives, formulated through the GFAP Constitution, are met. As successive commentators have observed, the Bonn Powers have created a perverse incentive for political parties in BiH to withdraw from passing legislation that would be contentious to electorates on the basis that the High Representative will pass it on their behalf. Through this situation we can see that legal authority in BiH is not simply devolved to the two Entities and Brčko District, but is rather derived from international law, supported through the Peace Implementation Council (PIC) and the UN Security Council.

The reliance on an international legal arena is evident in a second aspect of the accountability gap: the judicial response to crimes committed during the war in 1992–95. The use of the ICTY as the primary organ of transitional justice over the first decade after Dayton had profound social and political effects in BiH. Perhaps most starkly, through the indictment of just 161 individuals it focused attention on a small number of high-profile perpetrators of war crimes. As Dragan Musić from the Human Rights Centre in Sarajevo ruefully noted, the ICTY 'did not deal with the people who were responsible for the crimes, perhaps certain individuals were removed, but not all'.[1] It is telling that Dragan uses the phrase 'removed', since this relates to the second aspect of the ICTY: it has involved the use of justice at a distance, where Bosnian political elites could excuse the absence of debate concerning the events of the war since this was being confronted 'elsewhere' (see Jeffrey, 2011). Thirdly, the use of the ICTY has led to a perception in some quarters that the enactment of transitional justice is an internationally led process, an imposition in the same mould as other legal and legislative frameworks.

These concepts of selectivity, separation and imposition should not be seen as benign side effects of the chosen approach to transitional justice or divorced from the wider struggle to establish a state court and prosecutor's office in BiH. Rather they reflect the specific history of the creation of the ICTY and Court and their insertion into the wider geopolitical framing of BiH as a specific form of problem (following on from the work of Campbell, 1998a). In particular, the contested process through which these judicial instruments have come into existence illustrates how concerns of conflict resolution based on ethno-national territorialization shaped understandings of justice and sovereignty following the conflict.

6.1.1   The Creation of the ICTY and the Court of Bosnia and Herzegovina

The creation of the ICTY was a significant legal innovation that emerged from international indecision over how to respond to the conflict in BiH (see Chapter Three; Campbell, 1998a; Ó Tuathail, 1996). By October 1992 it was clear that widespread human rights abuses had occurred, from the shelling of Sarajevo to the allegations of concentration camps run by those loyal to the causes of the RS. There was a clear disconnect between foreign policy discourses, emerging from Western Europe, of an unfathomable conflict which was without clear victim and aggressor, and the imagery of persecuted civilians appearing on television screens. As Ó Tuathail (1996, 2002) has argued, the events in BiH were difficult to codify in foreign policy scripts, as Western leaders struggled to find the terms to explain the excesses of violence witnessed through the summer of 1992. As discussed above, the primary means of international intervention were diplomatic attempts to resolve the violence by proposing a division of BiH's territory that would be acceptable to all warring factions.

But these diplomatic actions were taking place amid growing moral outrage, both within and beyond the borders of the BiH. Over the course of spring and summer 1992, refugees crossing out of BiH spoke of a systematic move to expel non-Serb populations from towns and villages across swathes of western and eastern BiH. Later that year, international outrage was sparked by the reports of journalists Ed Vulliamy, Penny Marshall and Ian Williams concerning Trnopolje concentration camp, complete with the now infamous imagery of emaciated men and boys behind barbed wire (for a discussion of this imagery see Campbell, 2002a, 2002b). The journalist Pierre Hazan (2004) documents the tension among the UN Security Council members in response to the existence of these accusations of war crimes. Public clamours for accountability in Europe and North America risked presenting chosen peace makers, such as Radovan Karadžić, as war criminals. As a concession to the allegations of violations of the Geneva Conventions, and following pressure from French and US delegations to the UN, in October 1992 the UN Security Council agreed to establish a commission to gather evidence in BiH. This body, known as The Commission of Experts Pursuant to Security Council Resolution 780 (abbreviated to 'The Commission of Experts') and headed by Egyptian law professor Cherif Bassiouni, was hampered from the outset by the lack of allocated funds. Pierre Hazan (2004) suggests that the lack of resources was the consequence of a concern among UN Security Council members that this inquiry could harm the ongoing quest to find a diplomatic solution to the violence. In these terms it constituted a performance of accountability without substance. However, this did not account for the social and cultural capital of Cherif Bassiouni. Frustrated at the lack of financial and logistical support at the UN offices in Geneva, Bassiouni shifted the Commission's activities to his university offices at DePaul University, Chicago, in order to ensure computer access and establish a secure archive of testimonies and observations (Hazan, 2004: 26–30; Jeffrey, 2009b).

The case of The Commission of Experts serves as a precursor to the establishment of the ICTY. In 1993, US President Bill Clinton viewed the establishment of an international tribunal as an alternative to his election campaign pledge to lift the arms embargo on the former Yugoslavia and carry out air strikes on Serb military positions. This new political will was supported by the then French foreign minister Roland Dumas, who since 1992 had mounted a personal campaign to establish a war crimes tribunal in the face of continuing reports from BiH of systematic rape, executions and the expulsion of civilians from their homes (Hazan, 2004: 34–37; see Jeffrey, 2011). Despite continuing reluctance from the UK, and concerns from China and Russia over precedent setting, the Security Council established the ICTY through Resolution 827 on 25 May 1993. In response to what it considered 'grave breaches of the Geneva Conventions', Resolution 827 tasked the Tribunal with 'prosecuting persons responsible for serious violations of international humanitarian law committed in the territory of the former Yugoslavia' from January 1991 onwards (United Nations Security Council, 1993).

The mandate of the ICTY states that it may claim 'primacy and may take over national investigations and proceedings at any stage if this proves to be in the interest of international justice' (ICTY, 2012). The Tribunal was comprised of three organs: a judiciary, initially consisting of eleven judges though by 2008 they numbered thirty; the Office of the Prosecutor, a position held by Richard Goldstone (1993–1995), Louise Arbour (1995–1999), Carla Del Ponte (1999–2008) and since 2008 by Serge Brammertz; and the Registry providing administrative support.[2] While parallels can be made to the mandate of the Nuremberg Trials, the ICTY differs in its attempt to foster a sense of legal parameters to war *during* the conflict as opposed to after its conclusion. While this is a significant legal innovation it almost proved to be the Tribunal's undoing. Wary of unsettling the diplomatic efforts to resolve the conflict, Richard Goldstone was reticent to indict key protagonists in the violence between 1993 and 1995. In the post-conflict period the numbers of indictments have increased, and to date the ICTY has indicted 161 individuals on charges of violations of the laws or customs of war and breaches of the Geneva Conventions.

From its initial design within Resolution 827 the ICTY is conceived as an *ad hoc*, and therefore temporary, judicial institution. Part of its 'completion mandate' comprises the devolution of war crimes cases and prosecution competences to national state courts. In August 2003 the UN Security Council passed Resolution 1503, which formally called upon national judiciaries in the former Yugoslavia to try cases against what were termed 'low and mid-level war crimes perpetrators' (ICTY, 2006). Over the preceding five years successive High Representatives had attempted to create a judicial institution in BiH capable of meeting such a demand. This process of establishing a court and prosecution service had been fraught, as initial collaboration in 2000 between the two Entity governments and the OHR to draw up the necessary Law on the State Court of Bosnia and Herzegovina had collapsed. Consequently Wolfgang Petrisch, then High Representative, imposed the Law in November 2000. There followed a period of intense political lobbying for the establishment of the court in tandem with an ambitious programme of international fund raising. Following these initiatives two donor conferences were organized, first in The Hague (2003), where €15.7 million was pledged, and then in Brussels (2006), where donors pledged €7.7 million. By 2005 the Court was in a position to accept transferred cases from the ICTY. In tandem with the creation of the court a new state-level prosecution service, State Investigation and Prosecution Agency (SIPA), was also established.

At the Court's inauguration in 2005 Carla Del Ponte, then Chief Prosecutor for the ICTY, viewed the establishment of the court as an essential process in devolving and de-politicizing the judicial process in BiH:

> Allow me to stress only a couple of issues, which, in my view, will be at the core of domestic war crimes prosecutions. First, they must be victims oriented/ victims-driven and not to be seen as a process for the sake of the process, justice for the sake of justice. Secondly, they must be credible and seen as a-political/ de-politicized proceedings – which is of tremendous importance for the witnesses. And thirdly, they will have to fight against deeply embedded prejudices, public misconceptions, unsettled grievances and probably political interference. (Inauguration of the court, Chief Prosecutor Carla Del Ponte, 9 March 2005, in ICTY, 2005a)[3]

The rhetoric of Del Ponte points to the lofty ambitions placed on CBiH at its creation in 2005. This was a shift in both the geography and the sociology of transitional justice in BiH. In terms of geography, international policy makers in the UN, OHR and ICTY perceived the establishment of the State Court as a fundamental step in the consolidation of Bosnian state territoriality. For the first time since 1992 a single legal jurisdiction could cover the Bosnian political space. But the ambitions extend beyond unifying the state polity. The comments from Del Ponte illustrate the ambition for the new Court to exceed simply the practice of criminal trials ('the process') and instead to contribute to more abstract questions of victimhood, social healing and reconciliation. The desire for a 'victim-centred' approach suggests a revision of the ICTY approach, where the social consequences of the war crimes were often seen as beyond the remit of the legal process (Hodžić, 2010). Instead the CBiH was performing a function as a symbol of a new and centralized judicial process that would allow justice to be *seen* to be done by the BiH citizenry.

6.2   Contesting the State

In many respects the CBiH represents an improvised Bosnian state in microcosm. The Court is located in a refurbished JNA (*Jugoslovenska narodna armija*) barracks five kilometres from the centre of Sarajevo (see Figure 11). Funded by the European Union and the government of Japan, the CBiH contains eight courtrooms each constructed and equipped through further funding from multilateral and bilateral sources. A plaque stands outside each room to specify which European government provided the funding for its refurbishment, often with secondary plaques listing which organizations paid for furniture and equipment. Walking the corridors of the Court and sitting in on its trials, it appears a rare performance of unity across Bosnian territory, an institution that has jurisdiction over the entire territory of BiH. Working in tandem with the SIPA, defendants are drawn from across the Bosnian state, both through transfers from the ICTY and, increasingly, through domestic investigative processes. Symbolically the Court is a celebration of the Bosnian state: the blue and yellow flags of BiH adorn the external and internal spaces, while the Bosnian crest sits above each judicial panel (Figure 12).

But beyond these observations the creation and organization of the CBiH point to its more improvisational character. From the outset this body has been supported by international personnel within both the Prosecutor's Office and the Judiciary. In 2008, sixteen of the fifty-seven judges were recruited from beyond the borders of BiH. This proportion has decreased in the intervening years, and by 2011 there were only four international judges working at the court. The involvement of an international judiciary has been

Figure 11 The State Court of Bosnia and Herzegovina, Sarajevo
Source: Author's photograph

Figure 12 The Judge's Bench, Court Room Six, The CBiH, Sarajevo
Source: Author's photograph

a focus of domestic political opposition, in particular from current RS President Milorad Dodik's *Savez nezavisnih socijaldemokrata* (SNSD) party. Following complaints by Dodik's party the mandate of the international judiciary was not renewed at the end of 2009, leading to concerns that judicial panels within ongoing trials would need to be changed, perhaps leading to the restarting of a number of high-profile and long-term cases. However, the High Representative intervened and the mandate for the international judiciary was extended to the end of 2012 (see ICG, 2011).

But political opposition to the Court is not restricted to criticisms of international involvement. As stated above, the CBiH is located in a former JNA barracks outside the centre of Sarajevo. When I asked a Court official why this location had been chosen she laughed and said that 'all the good buildings in the centre of town had been taken by other government or international organizations'.[4] But the choice has become a focal point for a wider set of political objections to the existence of the Court. This stems from the specific history of this site within the conflict 1992– 95. During the Yugoslav period the barracks were named 'Viktor Bubanj', after a celebrated pilot in the *Narodnooslobodilačka vojska i partizanski odredi Jugoslavije* (People's Liberation Army and Partisan Detachments of Yugoslavia, Tito's army during the Second World War). In May 1992 the barracks were commandeered by the *Armija Republike Bosne i Hercegovine* (Army of the Bosnian Republic or ARBiH, loyal to the Bosnian government) and renamed 'Ramiz Salčin', commemorating an ARBiH soldier who fought in the siege of Sarajevo and died in 1992.

From this point on, there is little agreement on the usage and nature of this site. The area in which it is located was certainly highly contested and heavily mined during the conflict since it was on the frontline between *Vojska Republike Srpske* (the army of the RS) and ARBiH soldiers. By the end of the conflict the site was largely destroyed, leading one Public Information Official at the Court to describe its condition in the early 2000s as 'four walls and a mine field'.[5] But in this uncertainty over past usage competing historical narratives have thrived. It is worth exploring these contested spatial histories further since they underline the significance of place and commemoration to the process of claiming state sovereignty in BiH. Indeed, in many respects the struggles over the history of the court reflect the interweaving of claims to victimhood, moral virtue and spatial justice that characterize the political landscape of Dayton BiH.

The testimonies of witnesses and defendants at the ICTY suggest a range of opinions as to the previous usage of the Vikor Bubanj/Ramiz Salčin Barracks. A number of witness and defendant testimonies at the ICTY have suggested that the site was used for Serb prisoner internment; indeed, the convicted mass rapist Radovan Stanković attempted to use the history of the site to argue against the transfer of his case from The Hague to the CBiH. In a response to the trial judge's question about his psychological health Stanković linked the construction of the Court to the actions of the US in Iraq:

> I want to go there [BiH] to fight them with the facts and arguments and to break them down. But please, they want to turn Viktor Bubanj, which is a notorious execution ground for Serbs and a concentration camp, to turn that into a court and try me there. This reminds me of a statement by George Bush when he says the prison, the camp of Abu Ghraib, where crimes were committed, will be torn down to erase any site of crime and a new camp, new prison, Abu Ghraib will be built there for new crimes. They didn't even tear down this notorious camp in Viktor Bubanj; they just repainted the walls, the walls of the concentration camp that still bear the Serbian blood, that are still speckled with Serbian blood; and this is where I am to be tried. Serbian blood, the blood of innocent civilians, flowed there in rivers, rivers of blood flowed there, people who were taken from the Sarajevo …. (Radovan Stanković in ICTY, 2005b)

At this point the judge turned Stanković's microphone off. We obviously need to be careful not to grant too much credence to this account, though others do corroborate that the site was used for the internment of soldiers from the VRS. For example, a witness in the trial of Dragomir Milošević[6] described the nature of the internment at the camp:

> The Viktor Bubanj Barracks was a camp for Serbs, and on the third – on the second and third floors, that was where their army had its quarters. When we entered the Viktor Bubanj Barracks they broke my ribs, and I can show it to you if need be, you can see it quite well. Then they knocked out my front teeth and I was in a sorry state, and one of them approached me and stuck a knife point into my eye, here where you can see a scar. And when another guard told him not to do that, he said, let me beat the Chetnik. When I was able to beat my Alija in the prison in Foća, my president, why wouldn't I be able to beat a Chetnik? (Anonymous witness in the Dragomir Milošević trial, ICTY, 2007)

The claims made at the ICTY reflect a wider movement among RS politicians and NGOs to shape the commemoration of the Viktor Bubanj/Ramiz Salčin site. In September 2003 the *Saveza Logoraša RS* (RS Association of Camp Detainees) organized a march on the CBiH site to protest about the absence of any commemoration to the site's former use (see B92, 2003). The organizers sought permission to establish a plaque on the building, marking the site as a former 'death camp' where '2000 Serb civilians were killed, assaulted or tortured' (Superbosna, 2003). The Vice-President of *Saveza Logoraša RS*, Slavko Jović, sought to draw a moral equivalency between the desire among those loyal to Bosniak causes for a memorial at Srebrenica and the demands among pro-Serb groups for a plaque on the

Court building (ibid.). It is noteworthy that the *Saveza Logoraša RS* draws this parallel between the events in Srebrenica in 1995 and the 'death camp' at the Court, considering the long resistance RS politicians have mounted to the recognition of genocide in Srebrenica (OHR, 2010). But the claims made by witnesses and defendants at the ICTY and the *Saveza Logoraša RS* have only contributed to a wider public sentiment in the RS that the building was an inappropriate choice for the CBiH.

We need to be wary of translating the political discourse of victimhood unproblematically into a moral right to establish historical fact. One of the key sources of information concerning the events of the conflict is the archive of indictments at the ICTY (see Campbell, 1998a; Jeffrey, 2009b). These documents have drawn on a wide range of witness testimony and attendant material evidence to identify possible crimes against humanity and violations of the customs of war. The indictments do not support the suggested equivalency between the events in Srebrenica in 1995 and in Viktor Bubanj/Ramiz Salčin between 1992 and 1995. Where there are fifty-seven separate indictments relating to the events in Srebrenica, there is no single individual or group indicted for the alleged events at the former army barracks. In addition, where 6414 victims have been 'conclusively identified' at Srebrenica (OHR, 2010), at the time of writing no victims have been discovered as a consequence of the events at the Viktor Bubanj/Ramiz Salčin site.

Indeed, a series of statements made at the ICTY point to a different usage. A witness in the trial of Dragomir Milošević described a scene at the Vikor Bubanj barracks as an organized military facility and relatively unchanged since its usage by the JNA:

> That was the seat of the military court and the military prosecutor's office. That was also the place where the detention units, the office for the develop-ment of military maps, and there was a section of military police which partly catered to the military justice organs and partly to staff. (Witness at the trial of Slobodan Milošević in ICTY, 2005c)

Of course, it is unclear how much Milošević would have known about the events at the site following its use by the ARBiH and its name change to Ramiz Salčin. But testimonies exist that challenge the image of the site as an illegal concentration camp during the 1992–95 period. A representative from the BiH Camp Detainees Association, acting as a witness at the trial of Momčilo Krajišnik (a former senior figure in the SDS, sentenced to 20 years in prison in 2006), described a more conventional form of detention facility:

> The Viktor Bubanj Barracks, as far as I know, was a traditional type of jail. There were Serbs detained there, also 150 Bosniaks who had been charged with crimes or had refused the mobilization call. I was interested in the Viktor Bubanj Barracks because of a colleague, Jovočić. There were people of various

ethnicities detained there and detained for various reasons. That's what I know about it. (Representative from the BiH Camp Detainees Association, trial of Momčilo Krajišnik in ICTY, 2005d)

It is no surprise that the ICTY transcripts produce a set of conflicting accounts of this site in eastern Sarajevo. The testimonies from witnesses and defendants at the ICTY contest the usage of a wide range of military and civilian buildings during wartime BiH, often with the purpose of justifying their legitimacy as military targets or resisting the return of displaced populations (see Toal and Dahlman, 2011). These processes illustrate how the physical command over space during the conflict has been replaced with a discursive struggle to control the production of knowledge about the meaning of places in BiH.

The contestation over the Court's past has fed into political debates concerning its present and future. Carla Del Ponte's aspiration for the Court to be perceived as a honest arbiter has been frustrated by numerous political attacks on its neutrality. Milorad Dodik, first as Prime Minister of the RS and later as President, has mounted a series of staunch criticisms of the Court and the validity of its jurisdiction over the RS. Perhaps most controversially, in December 2008 Dodik claimed that he would not allow 'Muslim judges' to preside over cases involving RS citizens:

It is unacceptable for the RS that Muslim judges try us and throw out complaints that are legally founded. And we think that it is only because they are Muslims, Bosniaks and that they have a negative orientation towards the RS, and we see the conspiracy that has been created. (Milorad Dodik in B92, 2008)

The clear intent by Dodik is to stratify the judiciary along religious, and hence ethnic, lines. The majority of cases brought to the Court have involved former VRS or JNA soldiers loyal to the causes of the RS, reflecting the greater number of recorded war crimes committed by these groups in BiH. But by implanting an ethnic matrix over the work of the judiciary, Dodik attempts to present this outcome as a consequence of underlying partiality among the Court and the Prosecutor's Office.

But political attempts to discredit the Court have not simply been focused on a supposed pro-Bosniak bias. In tandem with the complaints over religious affiliation of the judiciary there have also been attempts by RS political leaders to present the Court as a foreign and anti-democratic imposition. In April 2011 the RS National Assembly called a referendum on whether voters 'support laws imposed by the High Representative in BiH, in particular the laws on BiH's state court and prosecution' (see ICG, 2011: 6; *Nezavisne novine*, 2011). The referendum was adopted by sixty-six votes to ten in the RS Assembly, with the vote to take place within forty-five days. However, under pressure from the then High Representative, Valentin Inzko, and the EU

High Representative for Foreign Affairs and Security Policy, Catherine Aston, Dodik withdrew plans for the referendum (see European Forum, 2011). Despite this withdrawal, the attempts to undermine the CBiH reflect the wider public opinion in the RS towards the CBiH. The RS-based newspaper *Nezavisne novine* ran a poll in March 2011 that found that 49.1% of the public in the RS support the abolition of the Court while only 3.8% hold this opinion in the Federation (Vukić, 2011).[7]

The disputes around, first, the location of the CBiH and, second, its judicial mandate illustrate established lines of contestation over state practices in BiH. As we have seen in previous chapters, claims to victimhood and the anti-democratic nature of intervention are established themes of political discourse in contemporary BiH, each framed within the narratives of ethnic enmity. But we must be careful to avoid reifying these discourses as inevitable outcomes to attempts to establish a state-level judicial entity. We see in the practice of the CBiH attempts to build its legitimacy among the Bosnian population through processes of 'public outreach'. Such initiatives are becoming a common part of new institutions of transitional justice (Lambourne, 2012). In order to promote such initiatives the CBiH was established with a separate department dedicated to public information and outreach services. Through this programme we can again see attempts to enrol civil society institutions into the running of this state agency. But perhaps more significantly, the public outreach strategies are serving to create new conceptualizations of justice in BiH while shaping the activities of the court itself.

6.3   Enrolling Civil Society

Over the first six years of the ICTY's existence (1993–99) there was no formal public outreach programme. Through this period the ICTY placed an emphasis on the processing of trials and issuing indictments. The prominence of the legal process illustrates a point made by criminologist Kieren McEvoy (2007) that the practice and study of transitional justice has been characterized by an innate 'legalism': prioritizing legal institutions and knowledge over wider social understandings of justice (see Jeffrey, 2011). Despite the absence of an ICTY-sponsored programme, a range of human rights NGOs in BiH sought to perform this function over the late 1990s. In an interview with Danijela, a representative from Helsinški Komitet working in the northern Bosnian town of Bijeljina, she explained one of the key motivations in becoming involved with this work:

> What we recognized was that the biggest problem was the denial of the crimes committed there. And what was visible in Foća, for instance, when someone is accused of systematic rape of young women between the age of 12

and 30 years of age and someone from the Serb community says 'no this has not happened'. And this is when the ICTY has absolutely, beyond reasonable doubt established the facts on this case, and even the guy – it was [Dragoljub] Kunarac – he said before the Court that he did it and that he would do it again. He said that he was aware of what he was doing. [...] So what was happening at the ICTY was not having any effect on the local communities in BiH. And that was something we recognized as a big problem. (Interview with Danijela, Helsinški Komitet Bijeljina, 18 October 2009)

Danijela's comments raise a series of questions regarding the purpose and audience of the ICTY trial process. While there is a tendency within scholarly literature to laud the archive of testimony produced through the ICTY trials (see Campbell, 1998a; Jeffrey, 2009b), the absence of a formal outreach programme meant that this knowledge was not being communicated to victim and veteran communities in BiH. Indeed, the nature of the ICTY as an international, and distant, form of justice allowed local political elites to discredit its practices and allude to embedded anti-Serb (in this case) bias.

In contrast to the approach of the ICTY, from the outset in 2005 the CBiH prioritized public outreach programmes. The creation of the Public Information and Outreach Service (PIOS) within the Court ensured that there was a permanent point of contact for victims and potential witnesses, and this service was responsible for producing press releases on Court activities and summarizing trial processes on a weekly basis. According to one of the founders of the PIOS within the Court, their motivation for the creation of this service extended beyond a simple bureaucratic initiative and instead pointed to a different philosophy of transitional justice:

War crimes are not an issue owned by judges and prosecutors. It is an issue that is owned by the affected community. They have to deal with it over a long period of time, and we wanted to empower the community and so they would feel part of the Court. (ICTY Liaison Officer and co-instigator of the CSN, Sarajevo, 9 October 2009)

This statement illustrates a fundamental rethinking of transitional justice from the ICTY to the CBiH. Where the ICTY constituted a significant legal innovation as international law was institutionalized within an *ad hoc* tribunal, the CBiH was attempting the opposite: to create a 'normal' state court that was perceived as a credible legal institution by the Bosnian citizenry. This apparently banal objective poses a significant challenge. The CBiH needs to establish its legal credibility (to elevate itself above the political machinations of Bosnian society) while simultaneously attempting to stimulate public interest in the Court's actions (to ground itself in the lives of the Bosnian citizenry). This tension between legal abstraction and public participation shapes the ambiguities around the language of public outreach. The very term 'outreach' risks solidifying a public perception of a legal centre of expertise removed from the wider public periphery (see Lambourne, 2012). While Zlatan urges the participation of the Bosnian citizenry in the activity of the Court he is, through a language of community and empowerment, reinforcing the notion of the state court and society as separate categories that may be the sites for specific interventions.

The tension between legal abstraction and public participation is illustrated in the case of the CSN, which was established by PIOS in 2005 with a mission 'to integrate the mission of the Court into the wider Bosnian community'.[8] As we saw in the case of state building in Brčko District (Chapters Four and Five), PIOS looked to NGOs as the main instruments through which to achieve this objective. Of course, the demonstrations by *Saveza logoraša RS* in 2003 indicate that the work of the CBiH had come to the attention of NGOs, though not in the supportive capacity envisaged by PIOS.[9]

Over the course of 2006, PIOS enrolled five human rights NGOs from across the territory of BiH into the CSN: Žene ženama (Sarajevo), *Centri civilnih inicijativa* (or CCI, Mostar), *Izvor* (Prijedor), Helsinški komitet (Bijeljina) and Forum građana (Tuzla). The organizations were selected for their expertise in the fields of human rights and reconciliation, and the objective of the CSN was to create a sustainable network that would provide information about the Court, and in particular the WCC, across BiH.[10] The original CSN programme was funded by the Court for six months and provided resources to each NGO to dedicate two staff to work on networking activities. One of the key objectives behind the CSN was for the organizations involved to develop their own network with smaller organizations and MZs, and thus the reach of the Court would extend into Bosnian communities through the operation of civil society groups. This six-month pilot project was not repeated, in part because elements of the CSN had become self-sustaining (i.e. the forms of networking envisaged by the Court were part of the established practices of the organizations involved) and in part on account of the changing priorities of PIOS away from outreach work and towards public and media relations.[11] The shift in funding saw a number of NGOs within the CSN move away from involvement with the Court, a move that was evidence for one PIOS Official of the 'passivity' of some NGOs in BiH.[12] Rather than funding the NGOs directly, the Court provided each organization in the CSN with a letter of support which could be used to gain funding from elsewhere. The hope was that the social capital of connections with the Court might be commodified through the use of the supportive letter.

As part of a larger research project into the nature of public outreach programmes I interviewed a series of NGOs working in the field of human rights and reconciliation, and two such organizations were members of the CSN: Helsinški Komitet in Bijeljina and CCI in Mostar.[13] Both of these NGOs had been working in their communities since the conflict and had established strong links with community associations, with international officials within the OHR and government embassies, and with local state officials. Representatives from both organizations seemed to divide their relationship with the Court into two time periods: an initial phase when the CSN was being established and a later phase where contact with the Court became more sporadic and alternatives to the CSN model were adopted. In relation to the first phase, the representative from CCI talked of the model of interaction promoted by the public outreach programme:

> In the beginning the whole CSN was built on approaching [...] NGOs and making a local CSN in every one of the regions. And the idea was that we go to every municipality and find people who will be part of the CSN and that idea was that we approached the local support, the Centre for Social Work and some other people who are [...] mostly known in that community even if they are from sport, or cultural associations or whatever area. And the idea was that this small network in that municipality gives support to witnesses in this small community. (Interview with Sanela, CCI, Mostar, 12 October 2009)

At the heart of these discussions was the concept of being 'known'. There was a particular kind of knowing, however. According to the NGO representatives, the lack of ICTY involvement in public outreach programmes had seen a growing resentment and lack of trust among witnesses, victims and associated groups. The ICTY was known, then, but not trusted. Despite attempts to establish the network through brokers within the local community, CCI experienced a general suspicion among the BiH public as to the ability of the CBiH to make any meaningful difference to the absence of accountability for crimes of the past:

> We had a lot of problems. People don't trust the local police. People don't trust the Centre for Social Work. People don't trust anybody. I mean after so many years giving their testimonies for nothing, in the end they think it is not worth it. (Interview with Sanela, CCI, Mostar, 12 October 2009)

For the CCI the most productive approach to gaining trust was to reformulate the approach to engaging with victims, witnesses and veterans. Sanela was critical of the former ICTY approach of 'turning up in a suit and issuing a subpoena' and instead cultivated an approach formulated around an understanding of the wider social context within which the process of witnessing or victimhood was embedded. Bearing witness and giving testimony, for CCI, was a social practice that came only after other problems had been addressed:

> The original idea was that if for instance someone who is a potential witness who needs help, any help, we have someone in this huge CSN which could approve the help for that person. So [in] some cases people have economic

problems, they have social problems, and they need help in this before they decide to become witnesses and after they decide to become witnesses. (Interview with Sanela, CCI, Mostar, 12 October 2009)

This is a different understanding of the individual that gives testimony. Where the ICTY approach could be criticized for perceiving the witness as an instrument within the judicial process, CCI were attempting to engage with witnesses as members of a community who are suffering from the triple challenge of economic deprivation, social fragmentation and psychological trauma. Embedding the consequences of war crimes in this way challenged the categories used to assign guilt or victimhood. For example, at one point in the interview Sanela gave a sigh and said, 'the biggest problem is that we are traumatized, we are all victims'.[14] Sanela's point here was not to deny the existence of victims and perpetrators of war crimes – she is strongly committed to the legal process at the CBiH – but to illustrate the problem of looking only to these individuals as suffering from the ongoing effects of the violence of the 1990s. She felt that one of the consequences of prioritizing the legal process in this way was that the wider social consequences of the violence were omitted from political debate. Indeed, as we have seen, the devolved structure of the Dayton state has allowed parallel historical narratives of the conflict to thrive.

While the debates involved in CCI's approach involve challenging legalistic approaches to transitional justice, their resolution often centred on more mundane questions of the citizenship status of the individual concerned. One activity undertaken by CCI was informing witnesses of their potential eligibility to receive a small state pension on behalf of deceased relatives who were recognized as Civil Victims of War.[15] Sanela explained the struggle in convincing people that they were eligible for this money:

When we told them you have a right to this money, they would say 'no, who is going to give us money?' You know, people didn't trust anything. Even if you explain to them the law and their rights, they are sceptical about many things. (Interview with Sanela, CCI, Mostar, 12 October 2009)

One of the key reasons given during the interviews for the lack of trust in state and international institutions has been the tolerance of corruption and lack of progress on confronting the perpetrators of the violence. Thus the accountability gap discussed above assumes a social form by undermining the legitimacy of formal political structures. Such an absence of trust also points to the forms of politics that have thrived in the post-conflict period in BiH. As individual political connections with the state are loosened, so solace is sought in other, often ethnically constructed, forms of collective identity.

The capture of discourses of transitional justice by nationalist politicians and activists was a particular concern for Helsinški Komitet, the CSN participant organization in Bijeljina. As discussed above, this organization had a long history of working in public outreach, working first with the ICTY in 1999 and then joining the CSN in 2006. In an interview, one of its project managers, Gordana, stated that the central purposes for the CSN were, first, confronting the propaganda produced by nationalist politicians and, second, providing a means of truth-telling. One of her great regrets was the time taken to establish a public outreach scheme at the ICTY, time that allowed nationalist party leaders to discredit legal processes as politically biased and/or the product of international political agendas.

One case that was repeatedly cited during the interview was that of Radovan Stanković. Stanković was a member of the VRS who was sentenced in March 2007 to twenty years' imprisonment for the crimes of systematic rape and imprisonment of women in what was termed *Karamanova kuća* (Karaman's house) in Foča municipality during 1992. His case was the first to be transferred from the ICTY, despite his reluctance for his case to be moved from The Hague to the former barracks site (see above, p. 142). But in May 2007 Stanković escaped from prison, since he was sent to serve his sentence in Foča, of which he had previously claimed that it would be an easy place to escape because he had friends and contacts there (Balkan Investigative Reporting Network, 2007). This case illustrates the frailty of the Bosnian penal system, where despite a state-level prosecution and judicial system there is no state-level prison.[16] But it was not the institutional weakness that was the primary concern for Gordana. She saw this as another example of opportunities for the perpetrator to express their views while victims are silenced:

> In the RS there was no public space given to the presentation of facts. There is enough space given for Stanković to send letters from wherever he lives now, explaining that this is an attack on the Serb people and he is not guilty, and they gave him two pages in the newspapers. But no one is giving space to those who suffered. To the girl who is now 26 years old, or 27, and she is more dead than alive because she was kept for three months in Karaman's House and was systematically raped by more than twenty soldiers. (Interview with Gordana, Helsinški Komitet, Bijeljina, 18 October 2009)

In order to confront this situation Helsinški Komitet were engaged in a range of activities directed towards providing victims, witnesses and veterans with more information about the Court process. Such activities were structured around workshops, one day in length, where victims, veterans, witnesses and members of the CBiH would come together to discuss the outcome of a trial. The challenge for Gordana was bringing together groups that had previously been on different sides in the conflict to discuss a trial outcome. As in the case of CCI in Mostar, the social context was crucial and Gordana talked witheringly of organizations that sought to 'recruit' victims to a workshop or seminar by simply turning up in a particular community and expecting participation. Again, this reflected an instrumental approach to judicial processes, where the victim is expected to perform a particular function, an assigned role within a victim/perpetrator/witness drama. In contrast, Gordana emphasized the time taken to build these networks:

> But the biggest issue is actually the preparation work. It is a much bigger issue to get in the same place people from different ethnic communities who are perceived as victims. For example you have to spend three days drinking *rakija* [plum brandy] with [name] if you want him to come. Or in some villages you have to convince [name] to come to the Čelebići case. It just needs time. (Interview with Gordana, Helsinški Komitet, Bijeljina, 18 October 2009)

The provision of workshops and seminars allowed the Court to describe the trial processes (often using the original evidence) and explain the length and nature of the sentence. One of the main challenges Gordana faced in this last task was the widely held perception of lenience, in particular that perpetrators were granted short sentences, often in countries deemed desirable for emigration:

> We have this case before the ICTY of Dražen Erdemović who actually pled guilty, and said 'yes I am aware that I killed approximately eighty people' and then he receives five years of imprisonment and then a new identity – he is actually from Bijeljina – he and his wife and child are living in Norway now, a much better life than those who were victims of his crime, and this is something that people don't really understand. (Interview with Gordana, Helsinški Komitet, Bijeljina, 18 October 2009)

The value of the workshop is that it serves as a chance for jurists to explain the mitigating circumstances that shape the sentencing process. In this case Erdemović was given a ten-year sentence in 1996, reduced by the Appeal Chamber at the ICTY to five years (in 1998). The Appeal Chamber lists in its judgment a series of personal circumstances that were taken into consideration when issuing the more lenient sentence. There were, however, further mitigating circumstances used to limit his sentence:

> The Trial Chamber first considered [...] his age (*'he is now 26 years old ... he is reformable ...'*), his family situation (*'the accused has a wife, who is of different ethnic origin, and the couple have a young child who was born on 21 October 1994 ...'*), his background (*'... he was a mere footsoldier whose lack of commitment to any ethnic group in the conflict is demonstrated by the fact that he*

was by turns a reluctant participant' in the armed forces of the various parties to the conflict), and his character (*'the accused is of an honest disposition; this is supported by his confession and consistent admission of guilt ...'*). (ICTY, 1998; emphasis in the original)

The workshops granted an opportunity to explain the ways in which sentencing decisions are made. But they were not simply sites where the Court transmitted information to a passive audience. At times during the workshops participants would submit evidence of alleged war crimes, in the hope that these would be take on by SIPA. Gordana explained how this has occurred:

For example in Konjic after this event they [the BiH police] arrested a group of people based on the evidence emerging from this conference. The same issue happened in Brčko as well, Brčko was a case before the ICTY in case of crimes committed against non-Serbs. But a very strong Serb community organization gathered victims, they presented, well I am not sure how many kilos of documents they gave to the ICTY, related to crimes committed against Serb civilians in the area surrounding Brčko. And that case was transferred to the state court, to the state prosecutor's office and they are currently working on that. So actually, beside the fact that we have this kind of outreach, through these workshops in the local communities we have actually somehow initiated this cooperation between the community and the Court. (Interview with Gordana, Helsinški Komitet Bijeljina, 18 October 2009)

The activities of both CCI and Helsinški Komitet suggest a more deliberative concept of transitional justice fostered through the workshops, seminars and face-to-face interactions. This approach does not, of course, entirely bridge the imagined divide between law and society. For example, an ICTY public outreach official in Sarajevo felt that the judiciary in both The Hague and Sarajevo did not see any value in these activities, since if the law is applied appropriately 'there is nothing to discuss'.[17] But the analysis of the CSN illuminates the entanglement of the legal process in the lives of the BiH citizenry: the struggle to convince witnesses that there is a purpose to providing testimony or why victims should come forward with new cases. For the CSN the barriers to these processes were not legal. Rather, they sought to uncover the social, psychological and economic contexts that frame participation in law. Certainly the agendas of the CSN organizations were shaped by the priorities delivered from the Court, in particular over the course of the first six months of the CSN's operation. But just as the NGOs in Brčko flexibly deployed particular donor discourses in order to secure funding, while avoiding substantive changes to their practices, we see a similar process at work in the CSN organizations. The deliberations over transitional justice held through

CCI or Helsinški Komitet activities are often highly critical of the work of the ICTY and, less so, the CBiH.

This finding is reminiscent of Tania Murray Li's (1999: 289) observation of the 'fragile and contingent accomplishments' of development programmes in post-colonial Indonesia. Certainly we see in the CSN a form of compromised transitional justice, sitting between the reconciliatory approaches advanced in sites such as South Africa or the traditional *Gacaca* Courts enacted in Rwanda. From a critical standpoint we can see the CSN as another example of the intervening agencies in BiH attempting to wield civil society as an instrument, a means through which popular participation is fostered without challenging the idea of the Bosnian state. But the CSN organizations are not marionettes, and the discussion of their practices demonstrates the ways in which they are attempting to challenge official discourses of transitional justice and institute alternative practices and understandings.

6.4   Conclusion

In *Bosnia: Faking Democracy after Dayton* (1999), David Chandler critiques the forms of representative politics in BiH as illusory, a form of performance that appears 'as if' democracy is being cultivated while in reality forms of supervision entrench international authority. This argument raises a number of challenging questions as to what constitutes fake and authentic in post-GFAP BiH, and risks falling into a form of local determinism where authenticity is equated with the presence of Bosnian staff (irrespective of the forms of politics that are cultivated). Criticisms of the CBiH often follow similar lines: that this is a fake legal institution where trials are in process but they are merely a piece of theatre that is directed by (visible and invisible) international agents. The choice of site for the court and the requirement for OHR imposition of its legal mandate are used to support this argument. But such an approach neglects to place the court and its activities in the wider spatial and political context of BiH. Part of the challenge is the innate 'legalism' of studies of transitional justice, where the consideration of social or political factors is presented as polluting the purity of legal reflection (McEvoy, 2007). This chapter challenges such a legalist perspective to argue that it is through an engagement with the improvised character of judicial processes that its value, albeit fragile and contingent, to processes of state building and reconciliation come to light.

The establishment of the War Crimes Chamber of the CBiH has followed many of the established narratives of the improvised state developed over previous chapters. At first glance this appears a rare moment of unity in the fragmented institutional landscape of BiH, a state institution that has jurisdiction across both Entities. But on closer inspection, and by studying its practices, we see the forms of international intervention that are required to sustain its operation, and the consequent political objections posed by domestic political actors. Perhaps reflecting concerns over the fragility of the CBiH, in 2011 the OSCE proposed the establishment of a Bosnian Supreme Court, in order to respond to criticisms over political interference with the existing judicial institutions (see OSCE, 2011). But this chapter has demonstrated a range of ways in which the Court has attempted to build its legitimacy with the local communities in BiH, in particular through the establishment of a network of supportive civil society organizations. This material illustrated two important facets of the relationship between courts and the communities they serve. First, and perhaps most significantly, the legal processes in the court are often reliant upon collaboration with a range of associations and NGOs in order to complete the trial process. Reaching victims, establishing investigations and securing the participation of witnesses each required the assistance of community associations and NGOs. This reliance challenges the imagined binary between legal activities and the wider social context. Second, the process of public outreach was also, in small ways, shaping the forms of legal trial that took place, as new investigations were prompted following seminars and workshops on existing criminal prosecutions.

Despite the forms of participation fostered through the CSN, this cannot claim to be 'civil society-led'; its origins in the Court's outreach programme ensure that it is consistently presented by participants and outside observers as a 'top-down' initiative. There is evidence of alternative mechanisms of transitional justice emerging in the former Yugoslavia, though significantly these are not tied to single states but are rather transnational attempts to cultivate shared approaches to seeking truth and reconciliation. The most high-profile such attempt is The Regional Commission for Establishing the Facts about War Crimes and Other Gross Violations of Human Rights Committed on the Territory of the former Yugoslavia (or RECOM), a civil society initiative established in 2004 involving a network of 1,900 NGOs, victims' and veterans' associations, and individuals from across the former Yugoslavia. Over the past seven years RECOM has sought to build public and governmental support for a single truth-telling/seeking commission across the former Yugoslav states. This process has faced numerous barriers, not least states such as Slovenia and Croatia seeking to avoid being categorized with BiH and Serbia as sites of crimes committed during the fragmentation of Yugoslavia (see Jeffrey and Jakala, forthcoming). Again, we see the significance of state territoriality over the management of historical narratives, where processes of transitional justice are enrolled in the state-building objectives. In the following chapter I will explore how another international discourse has been utilized in the improvisation of the state: the incorporation of Bosnian into European structures.

Notes

1 Interview with Dragan Musić, Human Rights Centre, Sarajevo, 7 October 2009.

2 For a comprehensive overview of the structure of the ICTY see http://www.un.org/icty/glance-e/index.htm.

3 Permission has been granted by the UN for all excerpts from the ICTY transcripts; I am grateful to Kumiko Sugiyama at the UN Publications Board for help with this.

4 PIOS official, Sarajevo, 8 October 2009.

5 PIOS official, Sarajevo, 8 October 2009.

6 Dragomir Milošević was a commander in the VRS who was convicted and sentenced to thirty-three years of imprisonment in 2006 for war crimes conducted during the siege of Sarajevo, 1992–95. The comments in the following quotation are taken from his appeals trial hearing in 2007. The appeal reduced his sentence to twenty-nine years.

7 This was a poll of 1200 people (600 from each Entity) conducted between 25 March and 3 April 2011.

8 ICTY Liaison Officer and co-instigator of the CSN, Sarajevo, 9 October 2009.

9 Details on the funding of *Saveza logoraša RS* are difficult to establish, though the evidence from Brčko District (see Chapter Five) suggests that RS political parties and state associations are instrumental in the funding of politically compliant associations, particularly returns organizations, cultural associations and veterans' associations.

10 PIOS official, Sarajevo, 8 October 2009.

11 ICTY Liaison Officer and co-instigator of the CSN, Sarajevo, 9 October 2009.

12 PIOS official, Sarajevo, 8 October 2009.

13 The research on which this chapter is based took place across two research trips to BiH in January 2007 and October/November 2009. In addition to interviews with representatives from CCI in Mostar and Helsinški Komitet in Bijeljina, I also interviewed representatives from the ICTY public outreach division in Sarajevo, officials from PIOS in CBiH, representatives from the UNDP *Access to Justice* programme, the Council of Europe's media and civil society division and the Human Rights Centre in BiH. I also attended war crimes trials and interviewed NGO volunteers for smaller organizations working in Sarajevo, Brčko and Mostar.

14 Interview with Sanela, CCI, Mostar, 12 October 2009.

15 The legislation differs in the RS and the Federation, but both offer pensions for victims of war that suffer '60% or more' physical disability including death or disappearance (see FBiH Law on Principles of Social Protection, Protection of Civil Victims of War and Protection of Families with Children, article 54 Official Gazette of the FBiH no. 36/99 and The Law on Protection of Civil Victims of War, article 2 Official Gazette of RS, no. 25/93).

16 The CBiH has a very limited detention facility which can hold twenty-one prisoners. By 2009, seventy-eight suspects or indictees had been ordered into custody by the CBiH, which has to pay a daily fee to use Entity prisons. While a state prison has been much heralded (and foundation-stone-laying ceremonies conducted), it is (at the time of writing) yet to be built (see Husejnović, 2009).

17 ICTY Liaison Officer and co-instigator of the CSN, Sarajevo, 9 October 2009.

Chapter Seven

Becoming European

Over the past decade international agencies in BiH have sought to shift the rubric of intervention away from discussions of post-conflict 'emergency' powers and towards the conditionality attendant upon the state's attempt to accede to the European Union. For example, since 2002 the High Representative within the OHR has shared the role of EU Special Representative with a mandate to 'achieve progress in implementing the Dayton Peace Agreement as well as in the Stabilisation and Association Process' (European Union Special Representative, 2011). Infrastructure projects across BiH celebrate financial assistance from the EU, local police are assisted by the EU Police Mission, while security is guaranteed through the presence of a European military force, EU-For. But the prevalence of Europe is not restricted to visual referents on the physical landscape. The requirements and criteria of accession to the EU permeate political discourse in BiH, with each political party often claiming its right to being the 'party for Europe'.

This chapter will explore how ideas of state improvisation can contribute to understanding the current processes of Europeanization in BiH. This discussion explores how ideas of Europe have been variously deployed by international and domestic political actors to bolster a range of state strategies. Far from constituting a form of 'cosmopolitan' or 'de-territorialized' solidarity, many of the ideals of political community advocated through European discourses are based on territorial expressions of the state. There are a number of lines of conflict in this process. The first is between what is said and what is desired. The recourse to European rhetoric may, at first sight, appear to be a gesture to a transnational political order, but its implementation has entrenched territorial practices of the state. The second line of conflict relates to different interpretations of what 'being European'

The Improvised State: Sovereignty, Performance and Agency in Dayton Bosnia, First Edition. Alex Jeffrey.
© 2013 John Wiley & Sons, Ltd. Published 2013 by John Wiley & Sons, Ltd.

entails. A central schism exists between those who seek to portray European identity as a form of openness to plurality and difference (see Amin, 2002, 2004), and those who seek to convey 'being European' as a fixed set of cultural and religious traits. These performances of Europe depend upon different resource sets, variously drawing on the forms of law and authority embedded in the GFAP and, in later examples, the geopolitical stories of the 'fault line' conveyed through religious and cultural histories.

There is no single process of Europeanization; it is a term that invokes a series of different political and social processes. The roots of European integration are in peace building: the inception of a European project at the Treaty of Rome in 1947 was directed towards moderating the egotistical ambitions of individual states following the Second World War (Shore, 2000). As the EU has expanded, its institutions and structures have been increasingly called upon to adjudicate in complex territorial disputes, and the resolution of such disputes has often formed part of the conditionality of initiating EU accession (for example, in Northern Cyprus and Serbia and Montenegro) (see Coppieters *et al.*, 2004). Such progressive 'Europeanization' has included new democratic forms, where certain aspects of state sovereignty are ceded to the European level (through the pressure of conditionalities rather than direct instruments of government), while the state remains the central locus of political power.

Consequently, underpinning the process of European integration in BiH is a wider question of the political integration of sovereign states into transnational governmental entities. Necessarily these processes invoke a sense of cosmopolitan democracy, a concept made popular in the 1990s by scholars such as David Held and Daniele Archibugi (1995). Their work explored the philosophical and practical basis on which new democratic institutions could be forged at a level 'above' the state (see also Beck and Grande, 2007). In particular these scholars questioned the democratic deficit of contemporary state-based democracies, arguing that these political systems lacked the ability to intervene in a range of global issues, including environmental degradation, corporate-led globalization and forms of transnational terrorist activity. But this deficit is not restricted to issues or process; Held and Archibugi (1995) point to a range of institutions such as global financial markets, multinational corporations and banking institutions which increasingly act in unilateral ways that influence national policies and strategies (see also Shapiro and Hacker-Córden, 1999). The recent global focus on credit rating agencies and the political effects of their judgement illuminates these concerns (see Willke, 2010).

In contrast to state formations, Franceschet (2000) and Held and Archibugi (1995) call for overlapping non-territorial forms of democratic governance that would both restrict and complement the existence of sovereign states. Jürgen Habermas (2001: 6) conceives of Europe as such a cosmopolitan democratic form, suggesting that the challenge of Europe is not to '*invent* anything but to *conserve* the great democratic achievements of the European nation-state, beyond its own limits' (original emphasis; see also Nava, 2002; Velek, 2004). Europe, then, is perceived to offer a continental-level form of cosmopolitan belonging, cultivating forms of political participation and citizenship outside of narrow nationalist agendas.

The ideals of European integration need to be set in contrast to the experience of the actual existing accession process. Over the following three sections this chapter will examine how ideas of Europe and cosmopolitanism circulate in contemporary BiH. The first section examines how the OHR have used forms of conditionality to link the process of European accession with the strengthening of the GFAP constitution. These processes are reminiscent of the underlying attachment to 'barrier' geopolitics within the GFAP, where intervening agencies consider strengthening BiH state sovereignty as a mechanism through which forms of liberal democratic citizenship may be cultivated. The second section explores how local politicians in BiH have drawn on European imaginaries to support political causes that are divergent from cosmopolitan ideals. The analysis focuses on the political discourses of Serb-nationalist political parties and youth associations. Reworking Milica Bakić-Hayden's (1995) concept of 'nested Orientalism', this approach illustrates the 'nested Balkanism' within European aspirations, where political parties have been keen to isolate different cultural groups as 'un-European' and 'Balkan' in contrast to their own virtue and modernity. These ascriptions rely on rigid cultural categorizations of 'being European' and operate at a distance from cosmopolitan ideas of transnational political solidarity. The concluding section explores the implications of these discussions for understandings of the improvised state. Where European accession could be seen as the waning of state significance in BiH, the processes documented in this chapter illustrate the entanglement of European discourses with divergent performances of the state in BiH.

7.1   Europeanization and the State

[…] EU membership will lock this country firmly into the democratic mainstream. It means access to EU development funds that can help turn the economy around. It means more foreign investment, creating more jobs. It means European standard justice. It means that – in the run up to membership – Bosnian politicians will have to show common sense and legislate the huge number of laws that are required to bring Bosnia into line with European standards. Each of those laws will help initiate improvements in living standards.

Ashdown (2005)

In setting EU membership in these terms, Ashdown makes a connection between accession and the establishment of democratic norms and values. The close articulation between Europeanization and democratization is understandable given that within EU enlargement documentation Europe is presented as 'an area of freedom, security and justice' (Commission of the European Communities, 2004). Where the Bosnian state has failed to act as a locus of citizenship or democratization, Ashdown's invocations of supra-national sovereignty look beyond the nation-state to the protective and democratizing values of the EU. This rhetoric conjures an image of democratic cosmopolitanism, where membership of the EU establishes an accountable structure of governance 'above' the scale of the state (see Held and Archibugi, 1995). In contrast to OHR-led practices of Bosnian state building, where a large percentage of the population (predominantly Serb and Croat constituencies) did not consent to the project, there appears to be universal support from Bosnian political parties for integration into Europe (Commission of the European Communities, 2003; Hayden, 2002).

But this virtuous narrative of Europeanization, where increasing integration into European structures affords democratic opportunities for the Bosnian citizen, underplays the conflicts and contingencies that have shaped the implementation of this policy in BiH. With particular reference to the book's core argument relating to the improvisation of the state, in this section I will draw out two points that serve to problematize the invocation of a 'transition' from an imagined Balkan past to a European future. First, the process of 'Europeanization' has not significantly reconfigured the power relations of international intervention: the OHR's repeated references to 'European values' mask the differential power positions of the actors involved in this political negotiation, while the abstract claims to democratization pay little attention to meaningful participation at the local level. Second, despite rhetoric of democratization and cosmopolitan political values, the central political effect of closer integration with Europe has been the strengthening of the Bosnian state. These two points are explored below through an examination of conditionalities relating to Bosnian entrance to the Council of Europe and the opening of Stabilisation and Association Agreement (SAA) talks.

Until the opening of SAA talks in November 2005 the EU had no formal contractual relationship with BiH; their contact has thus been 'short, but intense' (Commission of the European Communities, 2003: 5). But despite the absence of formal obligations, the EU and BiH have been in 'structured dialogue' since the GFAP (see Commission of the European Communities, 2005). In recent years this dialogue has stimulated a number of high-profile contacts between the EU and BiH. For example, since March 2002 the High Representative (then Lord Paddy Ashdown) has simultaneously held the post of EU Special Representative, to form the central point of contact between the EU and BiH. The EU have, as stated in the introduction, also taken over other defence and security competences over the past decade, most notably with the EU Police Mission and the EU security force, EU-For (see Juncos, 2005).

But to reduce the role of Europe to these tangible aspects would be to overlook the patterns of influence and authority that European institutions have exercised in BiH since the GFAP. Part of this influence has been mobilized through the lengthy procedures to join the Council of Europe (CoE), an organization that, while not directly affiliated to the EU, seeks to monitor and harmonize social, governmental and legal structures across its forty-six member-states. In 2001 the CoE gave BiH a series of political, social and economic criteria as conditions for gaining membership of the group. The level of detail within this document indicates how the conditionality of CoE membership was closely embroiled with the objectives of the international supervision of BiH. In particular, the first criterion set out by the CoE is '[t]o co-operate fully and effectively in the implementation of the Dayton Peace Agreements, which notably require the settlement of internal and international disputes by peaceful means' (Council of Europe, 2001). While further criteria refer to the cooperation with the ICTY and the ratification of the European Convention for the Protection of Human Rights and Fundamental Freedoms (ECHR), other aspects articulate closely with the practices of the OHR. For example, criterion IV(c) states that the Bosnian government must 'adopt, within six months after its accession, if it has not yet been done, the laws which have been temporarily imposed by the High Representative' (Council of Europe, 2001). This presents the Bosnian interlocutors with an open-ended conditionality, where membership of the CoE is dependent upon the fulfilment of laws that are yet to be imposed by the OHR. This situation became tautologous when the then High Representative Wolfgang Petritsch placed pressure on the Bosnian House of Representatives in 2001 to adopt a new election law, since they were failing in their fulfilment of CoE conditions.

The OHR and CoE conditionalities are thus seemingly entangled, their combined instruments of authority urging the implementation of the GFAP while reproducing international authority. Following the adoption of a new election law in August 2001, BiH was successful in its accession to the CoE in April 2002, leading the High Representative Wolfgang Petritsch to celebrate that BiH had found a 'European perspective':

> None of the mainstream parties now dispute the central political tenet that integration in Europe is the overarching aspiration of politics, economy and society in Bosnia and Herzegovina. (OHR, 2002: np)

The penetration of the 'European aspiration' to the heart of political, economic and social life in BiH was acutely felt through the subsequent conditionalities attached to opening negotiations on the SAA. Like the CoE criteria, a 'road map' was produced for Bosnian accession to the EU, identifying eighteen steps necessary for the opening of negotiations on SAA. The EU deemed these initial steps 'substantially completed' in 2002, leading to a broader feasibility study for opening SAA talks. This study grouped the remaining objectives of SAA criteria under three headings: political criteria (democracy, the rule of law, compliance with the ICTY and human rights), economic criteria (fiscal sustainability, privatization and financial sector review) and criteria relating to the ability to assume the obligations of the SAA (covering issues of the implementation of reform, foreign policy and regional cooperation) (Commission of the European Communities, 2003). The primacy of compliance with the ICTY within this document has led Ó Tuathail (2005: 57) to remark that the 'the road to the EU runs through The Hague'.

The political and social priorities contained in the SAA feasibility study emerged from BiH's membership of the Stability Pact, an EU initiative established as a conflict-prevention measure 'aimed at strengthening the efforts of the countries of South East Europe in fostering peace, democracy, respect for human rights and economic prosperity' (Stability Pact, 2006). The resulting criteria for SAA differ from the CoE in that they purposely look beyond Dayton, acknowledging its flaws as a cumbersome and inefficient architecture of governance. In particular, the SAA criteria seek to dilute the primacy of ethnic identity with the territorialization of BiH through the strengthening of the state-level Council of Ministers, removing parallel functions at municipal, canton and Entity levels and strengthening a professionalized civil service (Commission of the European Communities, 2003). In doing so, SAA criteria have served a useful function for the OHR as a means of revising the Dayton constitution under the auspices of European integration.

While the OHR may enrol the powerful imagery and vocabulary of a decisive break from international supervision through Europeanization, the conditionality of CoE and SAA reforms seems to suggest significant continuities in the exercise of international authority in BiH. Thus I would suggest that three key points can be made in relation to emergent European rubrics in contemporary BiH. First, the deployment of Balkanistic rhetoric by international agencies (such as the OHR) has continued since the conflict, principally through the assertion that BiH is a state 'in transition' from a past of ancient hatreds to a new European future. Second, though the OHR have connected Europeanization and democratization, the discussion demonstrates that the conditionalities inherent in the process of Europeanization, through both the CoE and the EU, are intricately bound into the priorities and practices of the existing international agencies in BiH. When I met an assistant to the High Representative in Sarajevo in 2003, he spoke at length of the importance of European criteria in instigating state reform and integration, acting as a 'pull' factor, against the 'push' of the OHR.[1] This rhetoric echoes the oft-stated division between 'hard' Bonn Powers and the 'soft' conditionalities associated with membership of European frameworks. In practice the evidence presented in this discussion suggests that the distinction between these variants of international influence cannot be so cleanly delineated. Third, though bound in rubrics of cosmopolitan affiliation to a European citizenry, the conditionalities of SAA and CoE accession have been firmly rooted in the cultivation of strengthened state sovereignty and citizenship. The spatialities and chronologies of such geopolitical Balkanism can be usefully compared with the emerging European rubrics within BiH, where designations of 'European' and 'Balkan' are flexibly applied between opposing political groups. It is within such Balkanist scripts that radically oppositional concepts of Europe emerged. But despite diverging from the earlier narratives of Balkanist geopolitics, these concepts of 'Europeanization' retain an attachment to state sovereignty as the primary unit of political life.

7.2   Nested Balkanism

As discussed in Chapter Three, between 1992 and 1995 Serb paramilitary groups, supported by the JNA, carved the Republika Srpska from the Bosnian state as an exclusively Serb territory. The political underpinnings of such military and paramilitary actions emerged from the ultra-nationalist rhetoric of Radovan Karadžić, founder of the SDS, who outlined the exclusive spatiality of the RS through the blunt refrain that 'our territories are ours, we can go hungry but we shall remain on them' (Karadžić, 1991). Such a geographical imagination does not simply outline a set of spatial objectives, but simultaneously emphasizes the absolute nature of cultural difference within the political philosophy of the SDS. Echoing the *integralist* rhetoric of the French and British nationalist politicians studied in the work of Douglas Holmes (2000), it was 'heterogeneity' and 'rootlessness' that was perceived to pose a threat to Serb national interest in BiH. An SDS representative in Brčko alluded to this when he stated that the key failing of (the multi-ethnic) Brčko District was its heterogeneity, offering the explanation that 'we don't like being mixed, when there is mixing there are problems'.[2] This notion of 'mixing' relies on stable, knowable and essentially different ethnic groups comprising the key social and political cleavage in BiH.

The creation of the RS, then, was a process of 'un-mixing' the Bosnian population and creating an ethno-nationally homogeneous territory. The violence that accompanied this process was both physical and symbolic, from the expulsion of the non-Serb population through to the destruction of references to other ethno-national groups within the built environment. In the case of Brčko (discussed in Chapters Four and Five), its position in a strategic location connecting the two halves of the RS constituted a particular focus for Serb paramilitary action (see Kadrić, 1998). Such 'ethnic cleansing' continued in the post-conflict period in both the RS and parts of the Federation through policies passed at the Entity level designed to dissuade returns and solidify the gains of the war (see Coward, 2002; Dahlman and Ó Tuathail, 2005a, 2005b; Toal and Dahlman, 2011). As discussed in Chapter Four, from 1996 towns that had previously held a Bosniak majority within the RS, such as Brčko, underwent a rapid Serbianization, involving the renaming of streets, the construction of Serb-orientated memorials and the building of Serb Orthodox churches, often on the site of vacated Bosniak homes (ICG, 1998). The intention was to create an ethnically homogeneous state-like territory, while simultaneously removing the possibility of heterogeneous identities and affiliations.

The violence of the formation of the RS highlights the potential paradox of the current European preoccupations of Serb political parties. Over the past decade the manifestos of the main political parties in BiH have converged on the issue of Europe, each stating the 'overriding value of European integration' (UNDP, 2002: 4). In the case of Brčko, the political parties contesting the 2002 presidential election embedded their campaign materials in the language and symbolism of the European Union. For example, a billboard advertisement for the *Partija demokratskog progresa* (Party of Democratic Progress, or PDP), a moderate Serb-nationalist political party, declared their party's European credentials by exclaiming '*Да, Портале Европска а ортале Српска*' ('Yes, you can be European and you can be Serbian') (see Figure 13). The words are adorned with juxtaposed European and Serbian flags, and a picture depicting a woman standing over a child doing written work, under the phrase '*Да, Учимо*' (Yes, We Study'). Animating what Ó Tuathail refers to as the RS's 'existential crisis' (2005: 59), the wording of this advertisement appears to pose a direct challenge to the image of Serbian nationalism as parochial, traditional or depending on founding myths, and instead offering an alternative vision of a cosmopolitan Serbianism accommodated within the EU. It could be argued that rather than celebrating an established national space, this poster offers an anti-ontopological vision, one where solidarity does not rely on a particular fixed identity but rather on a shared modernity.

But an interpretation of the poster, and the political rubrics from which it emerges, as a performance of a 'new' Serbian political imagination ignores the extent to which such pronouncements of Europeanism are strategically relational. This point was clear in discussions with Serb political party members and representatives of Serbian civil society organizations in Brčko, where Serbian Europeanism was justified in *relation* to other non-European groups. 'You need to be realistic', said the founder of a Serbian Orthodox youth organization. 'Serbs are part of Europe, we have a Christian past.'[3] The idea of 'being realistic' was often used as a means through which nationalist viewpoints could be raised in the interview setting, presenting the opinion as common sense in

Figure 13   PDP election poster, Brčko, 2002
Source: Author's photograph, used by permission

comparison to the 'unnatural' nature of multi-ethnic BiH. In this register of cultural difference, Serbian claims to European membership stem from its religious heritage, a trait that sets them apart from the Bosniak community.

Thus a new terrain of Balkanism is opened where a Bosnian Serb claim to Europeanism is structured around the identification of a non-European other. Following Bakić-Hayden (1995: 922), this can be described as 'nested Balkanism', since 'the designation of "other" has been appropriated and manipulated by those who have themselves been designated as such in orientalist discourse'. This Balkanist ideology reflects arguments made in relation to the Battle of Kosovo Polje in 1389, where certain Serbian commentators and politicians have portrayed the battle as a defence of Europe (the Serbian Kingdom) against invading Ottoman troops (see Kalajić, 1995). Echoing strands of contemporary resistance to Turkish membership to the EU, this vision promotes European unity as a Christian affiliation rather than based on the spread of democratic principles of freedom and security. This directly challenges the rhetoric of CoE and SAA criteria, since these political requirements are structured around Bosnian state membership as a multi-ethnic polity, not on the membership of the Serb minority as part of a normative vision of Christian European identity.

While promoting the notion of a set of enduring cultural differences fragmenting Bosnian society, such nested Balkanism simultaneously serves to disrupt the chronology of the geopolitical imaginaries of intervening agencies in BiH. Rather than seeing European membership as a claim that is accredited through the recognition of certain criteria by international actors, the SDS representative criticized the process of European integration and simply stated that 'Serbs have a right to be part of Europe'.[4] Probed further, the representative of the SDS based this assertion of entitlement on the high culture of Serbian society reflecting its inherently civilized nature. Indeed, the central preoccupation of the three Serbian youth organizations in Brčko was the preservation of cultural heritage and 'developing spiritual identity',[5] through 'trips to monasteries',[6] 'youth discussion groups'[7] and a range of sporting activities. The conception of an enduring threat to Serbian cultural heritage articulated in these research encounters echoes a strand of contemporary Serbian victimhood, where notions of Serbian identity are mobilized as a means of explaining the marginalized position of Serbs within the European Union. In such accounts, Serbs are again the sole defenders of Europe, as they were in 1389, though this time from the secular and commercialized European values invading from the West (Čolović, 2002). These interpretations of European enlargement have redeployed Balkanist language to suggest that 'the shadow of the collapse [of Europe] began to spread the moment people in Western European countries lost their sense of real values, that is, when money, material concerns and economic interest took the place of philosophy, religion, history and politics' (Čolović, 2002: 39). This concept of a 'collapse' of European cultural values seems to reflect the assertions of Milan Kundera's *Tragedy of Central Europe* (1984) where he explores the disjunction between perceptions of 'Europeanness' across Central and Western Europe. Kundera outlines the irony of the cherishing of a 'European' cultural identity in then Communist Central Europe at a time when 'Europe' was no longer perceived as a cultural value in Western Europe. Through such tropes RS politicians can present intransigence in the face of requirements of the CoE or SAA as the 'authentic' defence of European values against the neoliberal interventions made in the name of the European Union (Kalinić, 2004), a stance that has found fertile ground in some strands of the academic left (see, for example, Johnstone, 2002). Concurring with the study of Holmes (2000), this political project appears to foreground the essential cultural difference of Serbs as a means of mediating the alienation of neoliberal reform. Within this optic, 'being European' is stripped of its cosmopolitan affiliations, and replaced with a parochial connection to the Serbian nation.

Mirroring the Balkanist geopolitics of the Bosnian war, this interpretation of Serbian Europeanness creates an idealized Serb (cultured and sacred) against a vilified European (vulgar and profane). But more than a judgement of character traits, this Balkanist binary has political effects. In shifting the debate to questions of essential identities, this register of Europeanization ignores the tangible political necessities of Bosnian accession, such as the reform of the Bosnian state. Indeed, this concept of Europeanism is structured around a competing state project, the defence of the sovereignty of the RS. This tension between the demands of European integration and the desire to retain the sovereignty of the RS has been demonstrated in the recent protracted negotiations over Bosnian police reform (see OHR, 2005; Wisler, 2007). In the case of Brčko, a number of NGOs felt that operating projects between the two entities (the Federation and the RS) was difficult owing to the lack of cooperation from RS authorities. This was evidenced by one youth NGO coordinator, who was responsible for five NGO projects across BiH operating on both sides of the Inter-Entity Boundary Line, who expressed frustration at the obstructive practices of RS officials towards reform of the Bosnian state:

> [...] there is not a willingness in RS to have projects on their territory that are governed by the state level, because it is seen as a weakening of the powers of RS. They [RS officials] have said to me, 'we are never going to accept the state system you know, the state level has been devised to allow the ethos of the Federation to have its power, and it will weaken the RS to support anything that gives the state level credibility, we would undermine the power of the RS. So we have to hold very tight to RS power and not give anything.' (Interview with youth NGO coordinator, Brčko, 7 May 2003)

Thus being European, within the optic of Serbian political parties, involves a defence of the RS against the erosion by international agencies seeking to strengthen the Bosnian state. Blurring sovereignty and cultural identity, this motivation to retain the distinctiveness of Serbian cultural heritage allows RS politicians to simultaneously announce European aspirations while defending the considerable powers of the RS. The evidence from Brčko would suggest caution in interpreting the circulation of European rhetoric within Serbian political parties as a shift to a more cosmopolitan ethos based on the spread of shared values. Rather, this discussion has challenged this image through a consideration of radical cultural Europeanism that does not promote a transnational belonging, but rather essentializes particular cultural traits as representing 'Europeanness'. In this way, political parties, such as the PDP, create a discursive space to promote Europeanism, while simultaneously blocking constitutional and institutional reform that would assist Bosnian accession to the EU.

7.3 Conclusion

One of the darkest moments of the conflict in Brčko was the destruction of the large nineteenth-century Hotel Posavina in the centre of town in April 1992. The hotel's popular coffee lounge and cinema were destroyed, leaving a charred shell overlooking the town's central square. The hotel was not targeted for its military threat, it was not used as a barracks and it held no strategic value within the geography of the conflict in Brčko. Rather the threat posed by the hotel was a cultural one: it symbolized the possibility of inter-ethnic exchange and heterogeneity. The international response to the Bosnian conflict was to subscribe to the central logic of such attacks, explaining the violence as a consequence of intractable cultural differences across the Bosnian state. The solution to the conflict, the creation of exclusive ethno-national territories in BiH, served to sustain this vision and created the conditions within which nationalist political parties could continue to thrive.

This chapter has used the analytical tools developed by critics of Balkanism to explore how performances of 'Europeanness' have relied on a simultaneous casting out of a non-European 'Other'. The chapter identified these practices in two arenas. The first explored the attempts by international agencies to position BiH as a state 'in transition' to European norms, a practice that serves to entrench a Balkanized imaginary of a state confined by its past and in need of expert assistance. But by constructing a purport-edly 'undemocratic' BiH, international agencies serve to recover an image of Western Europe as a symbol of democratic virtue. This dual identity formation accords with Slavoj Žižek's (1990: 50) assertion that it is in Eastern Europe that the West constructs its 'Ego-ideal', banishing the 'decay and crisis' of its own democratic practices and looking to the East 'for the authentic experience of "democratic invention"'. But analysis of interview and textual material draws into question the entanglement of 'Europeanization' and 'democratization'. Rather, the conditionality related to CoE and SAA negotiations suggests continuity in the mechanisms of international intervention and the reliance on building state sovereignty. In the second arena, the chapter explored how processes of 'Europeanization' have seen the adoption and redeployment of Balkanist imaginaries by nationalist political parties. This material brought to the fore the 'nested Balkanism' of a radical Serbian Europeanism, structured around essential cultural differences and founded on the rejection of Bosniak claims to a European heritage.

The mirrored discourses of 'Self' and 'Other' present in these two arenas of enquiry demonstrate the enduring flexibility and political force of label-ling social, cultural or political practices as 'European' or 'Balkan'. It is the central aim of this chapter to move beyond the identification of scripts of similarity and difference and to focus on their political effects. I have argued that ideas of Europe circulating in contemporary BiH do not challenge the primacy of the state, despite the prevalence of references to forms of solidarity beyond the nation-state. Rather, the virtue of European association has been deployed to legitimize the strengthening of competing visions of statehood in BiH. 'Europe', then, does not act as a marker of virtue, a sign of the benevolent intentions of international agencies or a radical break from the nationalist past of parties such as the SDS. Rather it is a discourse of occlusion, a term that serves to mask the political practice structured around struggles over state power.

Notes

1   Interview with assistant to the High Representative, Sarajevo, 28 May 2003.
2   Interview with SDS representative, Brčko, 14 April 2003.
3   Interview with the founder of the St Sava's Youth Association, Brčko, 3 December 2002.
4   Interview with SDS representative, Brčko, 14 April 2003.
5   Survey of the Serb Youth Association, Brčko, 21 October 2002.
6   Survey of the Grčica Youth Association, Brčko, 21 October 2002.
7   Interview with representative of the Serb Sisters' Association, Brčko, 23 October 2002.

Chapter Eight

Conclusion

In late November 2011 political leaders from BiH met in the Italian town of Cadenabbia on the shores of Lake Como to attempt to break the deadlock surrounding the reform of the BiH Constitution. Organized by the German policy consultancy Konrad Adenauer Stiftung, the meeting was also attended by representatives from the European Parliament, the European Commission and the European Court for Human Rights. It was the latest in a series of internationally sponsored meetings aimed at tackling the decentralization of the Dayton Constitution. Over the course of these discussions constitutional reform has been incentivized by the enticement of membership of the European Union. But progress in these negotiations has been limited, as the greater centralization of the RS (as compared to the Muslim-Croat Federation) allows its political leaders to push for more autonomy from BiH and even a two-speed approach to European integration. Reflecting this desire, at the Cadenabbia meeting RS President Milorad Dodik remarked:

> In the future, the BiH Constitution should contain a mechanism to allow participants to part peacefully and which would include a procedure to allow people to express their opinions about what they want to see accomplished. (Kaletović, 2011)

These comments underscore the open-ended nature of BiH constitutional discussions, where high-profile participants can casually suggest the fragmentation of the state, as if this is a legitimate democratic concern in a country that still bears the legacies of 'ethnic cleansing' (see Toal and Dahlman, 2011). The comments also illustrate how, over fifteen years after the GFAP, fundamental questions regarding the nature and

The Improvised State: Sovereignty, Performance and Agency in Dayton Bosnia, First Edition. Alex Jeffrey.
© 2013 John Wiley & Sons, Ltd. Published 2013 by John Wiley & Sons, Ltd.

organization of the BiH state remain open to dispute. In citing the need to allow people to 'express their opinion' Dodik is recalling the fundamental questions regarding the validity and scale of self-determination that characterized the initial independence movements fromYugoslavia in 1991 and 1992.

The process of constitutional negotiation is a stark illustration that the BiH state is an ongoing accomplishment, where the establishment of sovereign arrangements at the GFAP has required a perpetual process of reinscription, through forms of international governance, the establishment of new state agencies, new donor demands and incentives through the expansion of the EU and NATO. The passage of time has not equated with the centralization of the BiH state, rather the uneasy truce secured at Dayton has established the conditions for continued negotiation as to the most legitimate form of state within (and beyond) the territory of present-day BiH. In many ways these debates encapsulate the paradox of studying the state in general and the BiH state in particular: the language is suggestive of singularity and indivisibility (i.e. 'the state'), where as the practice is one of plural state projects coexisting uneasily within the territory and discourse of Dayton BiH.

In order to illuminate this paradox in this book I have argued for an understanding of the BiH state as improvised. Focusing on international intervention in BiH since 1995, the seven preceding chapters have sought to illustrate the existence of competing practices of the state based on different ideas of solidarity and space. Building on an emerging field of scholarly studies of the state, this work has sought to detach understandings of sovereignty from grand explanatory narratives of ethnic difference, incapacity to rule or fundamental deviance. Similarly, the argument has resisted reproducing an understanding of state building as simply a neo-colonial enterprise, where a centre of calculating power subjugates a vassal state in perfect knowledge of its exploitative rule. Instead, this discussion has focused on the political implications of state improvisation in shaping political allegiances and understandings. In particular, across Chapters Four to Seven the book explored the spatial expression of a series of concepts: democratization, civil society, justice and Europe. The purpose of this approach is to examine the emergence of the BiH state as a spatial and historical process; a complex set of practices and materials that have come together to attempt to express a set of competing ideas of sovereignty. In this final chapter I will sketch the main arguments across the previous seven chapters before outlining the key theoretical and empirical contributions of the book.

In Chapter One I outlined the significance of legitimacy to understandings of state power. Focusing on legitimacy brings to the fore the complex web of processes, materials and ideas through which the validity of certain ideas of the state is conveyed. Improvisation is a useful analytical lens through which to explore such legitimating practices, since it orientates attention on the relationship between what is possible and what is desired. Moments of political breakdown and reconstruction provide an opportunity to explore how new ideas of the state are communicated, remaining attentive to the circulation of alternative narratives of political community. The key point here is to understand how certain state ideas rise to prominence while others are cast out as illegitimate and illegal. The book has illustrated that this process of selection does not happen in theoretical isolation, but occurs through the interaction of individuals and forces in political and social life. Historical and geographical stories of difference play into this process and allow certain understandings of the state to be understood as 'natural' or 'inevitable'.

This is not to say that improvisation stands in contrast to alternative modes of political practices, for example calculation. Rather improvisation is used to explore the contingency and compromise that characterizes the implementation of state-building plans. As discussed in Chapter Two, improvisation includes a sense of planning, since improvisations are resourceful acts that are structured around previous learning and socialization. This concept of performed resourcefulness also draws attention to the ways in which presentations of the state require public demonstrations of legitimacy. This is perhaps particularly evident in cases of state building, where new institutions are attempting to communicate their authority to a range of audiences. The concept of improvisation, then, is one that embraces the limits to power, which demonstrates that agents cannot simply 'act as they like' since each act is dependent upon available resources, whether material or ideological.

Through the stories in this book we can see this process of improvisation on a number of scales. On a grand policy level we see the effects of previous political practices in the decision making during and after the conflict in BiH. As Chapter Three demonstrated, the interventions in the conflict in BiH drew on a range of histories that sought to explain conflict in the region in geographical terms. The fault line, the barrier and the vortex each legitimized different styles of intervention; and the subsequent chaotic and crisis-ridden international interventions undertaken to stem the violence in BiH between 1992 and 1995 illustrate the lack of consensus among intervening agents as to which geopolitical frame held most merit. At different times and contexts international and domestic politicians drew variously on these spatial stories to explain the inevitability of violence in BiH. In the GFAP we see how geopolitics provided the resources to legitimize a new and contradictory form of statehood, one that simultaneously endorsed the division of BiH into ethnically orientated territories and guaranteed the right to return of refugees and internally displaced people. This performance of the state, orchestrated by US negotiating teams led by Ambassador Richard Holbrooke, was directed towards audiences outside BiH, where the claim could be made that despite the conflict BiH had been retained as a multi-ethnic state. But inside BiH the reality of political paralysis, economic deterioration and emboldened nationalist politicians spoke of the compromises of peace.

Improvisation also illuminates a more intimate scale of state making. In Chapters Four and Five the discussion focused on the way in which ideas of the state have been conveyed in Brčko District. While this example differs from other regions in BiH in receiving greater international funds and attention, it offers a chance to examine the localized effects of international supervision on the practices and institutions of democratic politics. Chapter Four examined the range of state practices that comprised ephemeral political performances, but sought to convey stability and permanence. Drawing in particular on the resource of law, these improvisations embedded understandings of democracy and economy through the prioritization of civil society and neoliberal forms of economic transformation. But alongside these practices coexisted a range of alternative improvisations of state power. Examples such as the boycott of new District administrative structures by SDS politicians and the projecting of alternative understandings of place into the public sphere illustrate attempts to convey a coherent political space that has met with numerous challenges. Where discourses of intervention celebrate the democratization of political life, the process of intervention appears to illustrate greater centralization of political power with the supervisory system of the OHR in Brčko. The interviews with OHR officials illustrated a belief in the technical challenge posed by state building in BiH, where the establishment of centralized authority will naturally lead to greater public support for its activities. But the evidence within this chapter was suggestive of a different chronology, namely, that intervention was creating the conditions for continued supervision of a domestic political system precisely because difficult decisions around housing, schooling or electoral law were removed from political debate and implemented by the District Supervisor. As a number of respondents suggested, this form of state improvisation has created the political conditions for the leaders of nationalist political parties to present their policies as defending 'true' national interest against the undemocratic practices of the intervening agencies, notably the OHR.

Where Campbell (1998a, 1999) illustrates the role of Dayton cartography in supporting ethno-national understandings of space, this study explores how mechanisms of intervention after the signing of the GFAP have continued to bolster ethnopolitics (Mujkić, 2007). In this sense improvisation is more than simply a style of political or spatial practice; it reveals some of the mechanisms through which international intervention can strengthen the support base of nationalist political parties. By presenting democratization as a technical process that required certain administrative and institutional frameworks, intervention has removed many of the most contentious issues from the political arena. This process is reminiscent of

Chantal Mouffe's (2005) concerns over the absence of political choice within Western liberal democracies following the end of the Cold War, where political parties have coalesced around similar understandings of market-orientated progress. Mouffe argues that such a technical rendering of politics impedes the constitution of distinctive political identities. She continues:

> This in turn fosters disaffection towards political parties and discourages participation in the political process. Hence the growth of other collective identities around religious, nationalist or ethnic forms of identification. (Mouffe, 2005: 5)

We see, then, in the case of Brčko two divergent political outcomes from the District process. On the one hand the close supervision of the political, economic and cultural development of this municipality has provided the context for refugee return and multi-ethnic coexistence. The neutralization of aspects of the symbolic landscape and attempts to challenge nationalist commemoration of the conflict have provided an accommodating space that differs from the forms of sociality in other areas of BiH (see Dahlman and Ó Tuathail, 2006; Toal and Dahlman, 2011). But on the other hand, the forms of intensive intervention have denied political participation and strengthened ethno-national narratives that seek to brand intervention a neo-colonial enterprise. While such claims do not have a basis in the character and practices of the agencies involved, these arguments have allowed nationalist political parties to flourish in the post-GFAP period (see ICG, 2011).

The nature of political participation was further investigated in Chapter Five. While the previous chapter had explored how new ideas of the state are spatially realized, this chapter examined attempts to establish what can be termed, following Ferguson and Gupta (2002), the 'verticality' of new state agencies through the identification and funding of civil society organizations. Mirroring the experience of post-Socialist transitions in a range of Central and Eastern European settings, the process of democratization in BiH has centred on cultivating civil society groups as evidence of democratic participation. But such a generic understanding of the virtue of civil society ignores the specific histories and geographies within which these processes unfold. Looking again at the example of Brčko District, the chapter examined how understandings of the conflict in BiH as a 'humanitarian disaster' ensured that humanitarian NGOs performed a central function in the international response. In the post-GFAP period many of these organizations remained, reorientating their work to objectives of reconciliation or the promotion of human rights. Many of these activities appear at first sight to be unconnected to the practices of state building, since the organizations are notionally non-governmental and their activities directed towards universal human aspirations of tolerance and cooperation. But the evidence presented in Chapter Five is suggestive of financial and regulatory frameworks that shape the behaviour and attitudes of these organizations. Using Pierre Bourdieu's conceptual vocabulary, the chapter explored these interactions as exchanges of social and cultural capital, where individuals and institutions were attempting to achieve institutional survival by modifying their activities to reflect a range of political agendas. These relationships were not solely derived from international funding, since funding by RS institutions sought to cultivate religious and youth NGOs directed towards preserving Serbian cultural practices in Brčko District. It was suggested that both the international and RS approaches reflected the gentrification of civil society, where social space is rendered as a commodity that may be disciplined through the investment of capital. The distinction between state and society is thus performed through attempts to cultivate civil society. But the very process through which this cultivation takes place blurs the state–society binary, as state institutions are further embedded into social practices through regulatory and funding demands.

Following the exploration of Brčko District, Chapters Six and Seven examined improvisation at the level of the BiH state. The process of establishing the State Court of Bosnia and Herzegovina was explored in the first of the two chapters, a discussion that illuminated the significance of this new judicial institution for communicating the idea of a coherent BiH state. The devolved nature of criminal justice processes since the GFAP had forged a connection between locality and judicial redress, where local courts dealt with crimes and criminals could avoid prosecution by moving to a different part of BiH or (more effectively) leaving the country. But from the outset the establishment of the CBiH had more lofty ideals than simply providing an effective system of criminal justice. According to statements made by international officials at the time, it was designed as a symbol of state unity and a focal point for achieving a sense of justice concerning crimes committed both during the conflict and in the post-GFAP period. Mirroring the approaches taken in previous chapters, these ideals were then examined in practice, exploring this process of judicial innovation as a form of state improvisation.

The material and micro-geography of the CBiH provide a stark illustration of its improvised character. Establishing the Court relied on international officials using their political resources (namely Bonn Powers) to implement a new criminal justice act in 2002. But this move only created the conditions in which a court could be built; the establishment of the CBiH still required appropriate buildings and infrastructure. As discussed in the chapter, the location of the Court has proved highly controversial, as those loyal to Serb causes have sought to challenge the neutrality of its trial processes on the basis of its location within a former army barracks. While little agreement exists on whether the claims of past crimes in this space are correct, the debate over its past has proved a focal point for political critique. The challenges to the existence of the CBiH, and particularly its function of trying war crimes, have led court officials to explore mechanisms through which to strengthen the legitimacy of this organization with the BiH population. The chosen strategy has been to cultivate a series of NGOs through which members of community groups and local associations may be educated about the purpose of the Court and consequently build trust in its activities. While the evidence from research participants pointed to the benefits of this approach in sharing information about the CBiH, it also highlighted a more complex set of interactions that challenged the clear delineation between law and society. The activities of Court Support Network NGOs proved to be crucial to the enactment of the war crimes trial processes, since they facilitated the attendance of witnesses and provided details of new potential investigations. But these processes also illustrated the blurred line between formal legal processes and non-legal NGO activity. The establishment of the CBiH was designed to perform a coherent and authoritative state agency, but its operation has required building consent with a range of institutions distributed across the BiH state and beyond.

The final chapter examined whether discussions of state improvisation are relevant to processes of increasing European integration in BiH. The presentation of BiH as a non-European 'Other' often relies on Balkanist imaginaries, as explored in Chapter Three. With the accession of neighbouring states, BiH has found itself in a liminal space, neither part of, not apart from, the EU. The premise behind this discussion was the almost universal appeal of EU imagery and rhetoric for politicians across the political spectrum in BiH. This consensus suggests that the term 'European' has been hollowed out, only to be repopulated with a range of new significations. Certainly, the ideal of EU membership has provided members of intervening agencies with the opportunity to set a series of new conditions on EU membership, many relating to the renegotiation of the devolved GFAP Constitution. Rather than cultivating new forms of 'post-national' or 'cosmopolitan' citizenship, these activities illustrate the OHR's entanglement of European objectives with strengthening the capacity of the BiH state. Through this process the imagined 'hard' Bonn Powers are difficult to distinguish from the 'soft' power of EU conditionality. The chapter also examined how politicians from Serb parties have attempted to perform the delicate representational act of simultaneously endorsing both Serb national identity and the virtue of the European project. This has been achieved by presenting 'being European' as an essential set of cultural traits, relating in particular to religious identity. This approach recalls the fault line geopolitics of Chapter Three, where the primary form of political difference is traced back to a mythical defence of European values against the invading hordes of Ottoman forces at Kosovo in 1389. This narrative endorses a series of exclusive geographies, perhaps most explicitly directed towards preserving the state-like powers granted to the RS at the GFAP. Consequently the signifier of 'Europe' performs multiple political endorsements in contemporary BiH, torn as it is between competing state projects.

These arguments made across the seven chapters of the book carry a series of theoretical and empirical implications. In terms of theory they seek to advance an understanding of improvisation as performed resourcefulness, highlighting how understandings of statehood are articulated and practised by a range of socially embedded actors beyond those traditionally perceived as agents of the state. It allows insights into the ways in which agents subvert or rework ideas of the state, not as forms of resistance 'outside' state structures, but within the processes through which structures are reproduced. This account affirms a number of long-held positions within critical accounts of the state. It challenges the binary between state and society, instead exploring the simultaneous production of these categories through processes of state building. Second, and in conjunction, it explores the interplay and co-constitution of structure and agency. Like others (Giddens, 1984; Jessop, 1990, 2008), I am seeking to provide an account of social and political practice that charts a course between a suggestion of overbearing economic or political structure and that of an entirely unconstrained set of rational actors. Using the conceptual frameworks of Pierre Bourdieu we see how the circulation of capital can be used to understand how decisions are made through evaluations of the specific social and political circumstances of the group or individual in question. This is not an entirely flat hierarchy; the ability to create law and enact physical violence privileges certain actors, often international agencies or individuals, sometimes hard-line nationalist politicians, but more often (as we have seen) a complex entanglement of the two. Their will is not pursued through aggression or coercion but by setting a symbolic value on particular vocabularies, dispositions and activities. While this framework is helpful in explaining constraints on agency, it is also a useful way of illuminating how state performances are distributed across micro-situations and encounters.

But perhaps the single most significant aspect of the theoretical contribution of improvisation is its simple focus on practice. As stated in the introduction, there has been a shift in emphasis within recent political geography to understand geopolitical discourse as practice rather than text, as a form of meaning that is conveyed through materials, dispositions, enunciations and bodies (Dittmer and Dodds, 2008; Müller, 2008; Thrift, 2003). Improvisation has to unfold; it exists only in a dynamic form and, as theatrical accounts attest, by its very nature it is difficult to theorize as an abstract form. Contrary to the state as a unified actor or static form, understanding the state as improvised urges reflection on the unfolding of the state as a process always in motion. As Wendy Brown's (2010) recent account of the phenomenon of late-modern wall building illustrates, there is no vantage point from which a state will cease to evolve and transform its assertions of authority. The example of wall building illustrates the insecurity of purportedly 'secure' states (such as the United States) in having to perform their power through the theatricality of increased border security. Similarly, Chapter Six gives an account of the establishment of the State Court of Bosnia and Herzegovina: this is both a moment of state triumph (a long-awaited national judicial authority) and a signal of its fragility (the court had to be imposed by an international agency using supposedly short-term emergency powers eight years after the end of the conflict).

There are, however, limits to this approach. Improvisation has been adopted as a lens and not a metaphor; this style of analytics provides insights into practices and motivations but it does not produce a totalizing or predictive theory of the nature of states in general or the BiH state in particular. While the discussion illustrates a number of political and eco-nomic imperatives that shaped the symbolic value of particular practices, these are not reduced to a single logic of rule. Rather this approach returns the discussion to Timothy Mitchell's (1991) point, that understanding the state is both a methodological and a conceptual practice, where a state's legitimacy is traced through empirical engagements with the ability of particular understandings of sovereign power to change political and spatial practices. Necessarily these are fragmented and incomplete accounts which, at times, point to wider narratives of politics and economics, but often gesture back to more banal concerns of survival and respect in new political contexts.

This leads on to the empirical contribution of the book. The research has advocated a situated and grounded exploration of the production of discourses and practices of international intervention in BiH, exploring a process that has been unfolding since 1995 through four periods of residen-tial fieldwork since 2002. This strategy reflects the need to engage with qualitative methodologies in order to understand the state as an improvised set of practices. Exploring the practice of the state through extended interviews and participant observation captures the multiple sites and scales within which state-like practices unfold. When research is conducted over a series of periods of residential fieldwork the varied temporalities of state-building narratives also come to the fore, as geopolitical events and transformations lead to changes in international policy of intervention. Hence, a qualitative approach has merit beyond the example of BiH, as improvisation helps to illuminate the partial and plural nature of all state claims, grounded as they are in specific assertions of history and geography. These assertions may seem more explicit in cases of state building, as, in the case of BiH, the Cadenabbia meeting illuminates the uncertainties around the fundamental questions of the organization of structures of government. Similarly the currently unfolding discussion of the nature and organization of post-Gaddafi Libya illustrates the competing claims to state legitimacy that exist after periods of political transformation (see van Genugten, 2011). But all assertions of statehood are improvisations, since statehood is secured through repeated performances of power. That such improvisations do not appear impromptu and contested is an illustration of the unequal power positions of different states and their ability to present their actions as uncontested and timeless. The book started with the example of the Palestinian Authority attempting to claim a seat at the General Assembly of the UN. In this case state improvisation seemed explicit and *ad hoc*, but only on account of the lack of wider international recognition of their state claim. The lens of improvisation captures this fragile division between legitimate and illegitimate states, a distinction that contains no absolute or natural criteria but is rather asserted through performances of legitimacy. Using the starting point of improvisation ensures that it is this fragility and dynamism that remains the focus of scholarly scrutiny, illuminating the myriad and contested ways in which states attempt to reproduce their power.

References

Abrams, P. (2006 [1988]) 'Notes on the difficulty of studying the state', in A. Sharma and A. Gupta (eds) *The Anthropology of the State: A Reader*. Oxford: Blackwell, 112–130.

Agnew, J. (1994) 'The territorial trap: the geographical assumptions of international relation theory', *Review of International Political Economy* 1: 53–80.

Aitchison, A. (2011) *Making the Transition: International Intervention, State-Building and Criminal Justice Reform in Bosnia and Herzegovina*. Antwerp: Intersentia.

Allcock, J. (2000) *Explaining Yugoslavia*. London: Hurst & Co.

Allen, J. and Cochrane, A. (2007) 'Beyond the territorial fix: regional assemblages, politics and power', *Regional Studies* 41: 1161–1175.

Amin, A. (2002) 'Ethnicity and the multicultural city: living with diversity', *Environment and Planning A* 34: 959–980.

Amin, A. (2004) 'Multi-ethnicity and the idea of Europe', *Theory, Culture & Society* 21 (2): 1–24.

Anderson, B. (2006) 'Becoming and being hopeful: towards a theory of affect', *Environment and Planning D: Society and Space* 24: 733–752.

Andreas, P. (2004) 'The clandestine political economy of war and peace in Bosnia', *International Studies Quarterly* 48: 29–51.

Andreas, P. (2008) *Blue Helmets and Black Markets: The Business of Survival in the Siege of Sarajevo*. New York: Cornell University Press.

Andrić, I. (2000 [1941]) *The Days of the Consuls*. Belgrade: Dereta.

Anheier, H.K. (2004) *Civil Society: Measurement, Evaluation, Policy*. London: Earthscan.

Ansary, T. (2009) *Destiny Disrupted: A History of the World Through Islamic Eyes*. Philadelphia: Public Affairs.

Anzulović, B. (1999) *Heavenly Serbia: From Myth to Genocide*. New York: New York University Press.

The Improvised State: Sovereignty, Performance and Agency in Dayton Bosnia, First Edition. Alex Jeffrey.
© 2013 John Wiley & Sons, Ltd. Published 2013 by John Wiley & Sons, Ltd.

Ashdown, P. (2003) 'Peace stabilisation: the lessons from Bosnia and Herzegovina. Public lecture by Paddy Ashdown, European Special Representative and High Representative to Bosnia and Herzegovina, given at the London School of Economics on 8 December 2003.' *Discussion papers, DP27*. London: Centre for the Study of Global Governance, London School of Economics and Political Science.

Ashdown, P. (2005) RS Leaders Must Not Fail. http://www.ohr.int/ohr-ept/presso/pressa/default.asp?content_id=33913 (accessed 14 June 2009).

Azaryahu, M. (1996) 'The power of commemorative street names', *Environment and Planning D: Society and Space* 14 (3): 311–330.

B92 (2003) Spomen-ploča u bivšoj kasarni. 19 September. http://www.b92.net/info/vesti/index.php?yyyy=2003&mm=09&dd=19&nav_category=12&nav_id=119903 (accessed 13 August 2011).

B92 (2008) Dodik's statements stir new controversy. 12 December. http://www.b92.net/eng/news/region-article.php?yyyy=2008&mm=12&dd=12&nav_id=55686 (accessed 18 October 2011).

Bajrović, R. (2005) BiH municipalities and the EU: direct participation of citizens in policy-making at the local level. Sarajevo: Open Society Fund Bosnia and Herzegovina, 47 pp.

Baker, T. (2007) 'Resources in play: bricolage in the toy store(y)', *Journal of Business Venturing* 22 (5): 694–711.

Baker, T., Miner, A. and Eesley, D.T. (2003) 'Improvising firms: bricolage, account giving and improvisational competencies in the founding process', *Research Policy* 32: 255–276.

Bakić-Hayden, M. (1995) 'Nesting orientalisms: the case of former Yugoslavia', *Slavic Review* 54 (4): 917–931.

Balkan Investigative Reporting Network (2007) Stanković Still at Large. http://www.bim.ba/en/65/10/3139/ (accessed 20 October 2011).

Barnett, T. (2005) *The Pentagon's New Map: War and Peace in the Twenty-First Century*. New York: Berkley Books.

BBC (2011) Palestinians send homemade UN membership chair on tour. http://www.bbc.co.uk/news/world-middle-east-14787327 (accessed 20 October 2011).

Bebbington, A. (2004) 'Social capital and development studies 1: critique, debate, progress?', *Progress in Development Studies* 4 (4): 343–349.

Beck, U. and Grande, E. (2007) *Cosmopolitan Europe*. Cambridge: Polity.

Bell, D. and Binnie, J. (1994). 'All hyped up and no place to go', *Gender, Place. and Culture: A Journal of Feminist Geography* 1 (1), 31–47.

Bell, D. and Binnie, J. (2004) 'Authenticating queer space: citizenship, urbanism and governance', *Urban Studies* 41 (9): 1807–1820.

Belloni, R. (2001) Civil society and peacebuilding in Bosnia and Herzegovina, *Journal of Peace Research* 38: 163–180.

Bennett, C. (1995) *Yugoslavia's Bloody Collapse: Causes, Course and Consequences*. London: Hurst & Co.

Berliner, P. (1994) *Thinking in Jazz: The Infinite Art of Improvisation*. Chicago: University of Chicago Press.

Bernhard, M. (1996) 'Civil society after the first transition: dilemmas of post-communist democratization in Poland and beyond', *Communist and Post-Communist Studies* 29 (3): 309–330.

Bethlehem, D. and Weller, M. (1997) *The Yugoslav Crisis in International Law*. Cambridge: Cambridge University Press.

Bialasiewicz, L., Campbell, D., Elden, S., Graham, S., Jeffrey, A. and Williams, A. (2007) 'Performing security: the imaginative geographies of current US strategy', *Political Geography* 26: 405–422.

Bieber, F. (1999) 'The conflict in former Yugoslavia as a "fault line war"?', *Balkanologie* 3 (1): 33–48. http://balkanologie.revues.org/index283.html (accessed 17 May 2011).

Bieber, F. (2000) 'Bosnia-Herzegovina and Lebanon: historical lessons of two multi-religious states', *Third World Quarterly* 21 (2): 269–281.

Bieber, F. (2005) 'Local institutional engineering: a tale of two cities, Mostar and Brcko', *International Peacekeeping* 12 (3): 420–433.

Bilgin, P. and Morton, D. (2002) 'Historicising the representation of "failed states": beyond the cold war annexation of the social sciences', *Third World Quarterly* 23 (1): 55–80.

Bodnar, J. (1993) *Remaking America: Public Memory, Commemoration, and Patriotism in the Twentieth Century*. Princeton: Princeton University Press.

Bojičić, V. (1996) 'The disintegration of Yugoslavia: causes and consequences of dynamic inefficiency in semi-command economies', in D.A. Dyker and I. Vejvoda (eds) *Yugoslavia and After: A Study in Fragmentation, Despair and Rebirth*. Harlow: Longman, 28–47.

Bojičić-Dzelilović, V. (2003) 'Managing ethnic conflicts: democratic decentralisation in Bosnia-Herzegovina', in S. Bastian and R. Luckham (eds) *Can Democracy Be Designed? The Politics of Institutional Choice in Conflict-torn Societies*. New York: Zed Books, 277–302.

Bolton, M. and Jeffrey, A. (2008) 'The politics of NGO registration in international protectorates: the cases of Bosnia and Iraq', *Disasters* 32 (4): 586–608.

Bose, S. (2002) *Bosnia after Dayton: Nationalist Partition and International Intervention*. London: Hurst & Co.

Bougarel, X., Helms, E. and Duijzings, G. (eds) (2007) *The New Bosnian Mosaic: Identities, Memories and Moral Claims in a Post-War Society*. Aldershot: Ashgate.

Bourdieu, P. (1979) *Algeria 1960: The Disenchantment of the World, the Sense of Honour, the Kabyle House or the World Reversed*. Cambridge: Cambridge University Press.

Bourdieu, P. (1984) *Distinction: A Social Critique of the Judgement of Taste*. Cambridge, MA: Harvard University Press.

Bourdieu, P. (1986) 'The force of law: toward a sociology of the juridical field', *Hastings Law Review* 38: 805–853.

Bourdieu, P. (1987) 'What makes social class? On the theoretical and practical existence of groups', *Berkeley Journal of Sociology* 32: 1–18.

Bourdieu, P. (1989) 'Social space and symbolic power', *Sociological Theory* 7 (1): 14–25.

Bourdieu, P. (1990) *The Logic of Practice*. Palo Alto, CA: Stanford University Press.

Bourdieu, P. and Wacquant, L. (1992) *An Invitation to Reflexive Sociology*. Chicago: University of Chicago Press.

Bowman, I. (1928) *The New World Problems in Political Geography*. London: George C. Harrap & Co.

Bratsis, P. (2006) *Everyday Life and the State*. Boulder, CO: Paradigm.

Brčko Development Agency (2002) Development strategy of the Brčko District of Bosnia and Herzegovina 2002–2006. Brčko: Brčko Development Agency.

Bridge, G. (2001) 'Bourdieu, rational action and time-space strategy of gentrification', *Transactions of the Institute of British Geographers* 26: 205–216.

Bringa, T. (1995) *Being Muslim the Bosnian Way: Identity and Community in a Central Bosnian Village*. Princeton: Princeton University Press.

Brown, W. (2010) *Walled States, Waning Sovereignty*. New York: Zone Books.

Butler, J. (1990) *Gender Trouble: Feminism and the Subversion of Identity*. London: Routledge.

Butler, J. (1997) 'Gender trouble, feminist theory and pyschoanalytic discourse', in L. McDowell and J. Sharp (eds) *Space, Gender, Knowledge: Feminist Readings*. London: Arnold, 247–261.

Butler, J. (2004) *Precarious Life: The Powers of Mourning and Violence*. London: Verso.

Calhoun, C. (1993) 'Habitus, field and capital: the question of historical specificity', in C. Calhoun, E. LiPuma and M. Postone (eds) *Bourdieu Critical Perspectives*. Chicago: Chicago University Press, 61–88.

Campbell, D. (1998a) *National Deconstruction: Violence, Identity and Justice in Bosnia*. Minnesota: University of Minnesota Press.

Campbell, D. (1998b) 'MetaBosnia: narratives of the Bosnian war', *Review of International Studies* 24 (2): 261–281.

Campbell, D. (1999) 'Apartheid cartography: the political anthropology and spatial effects of international diplomacy in Bosnia', *Political Geography* 18 (4): 395–435.

Campbell, D. (2002a) 'Atrocity, memory, photography: imaging the concentration camps of Bosnia – the case of ITN versus *Living Marxism*, part 1', *Journal of Human Rights* 1(1): 1–33.

Campbell, D. (2002b) 'Atrocity, memory, photography: imaging the concentration camps of Bosnia – the case of ITN versus *Living Marxism*, part 2', *Journal of Human Rights* 1(2): 143–172.

Caplan, R. (2005) 'Who guards the guardians? International accountability in Bosnia', *International Peacekeeping* 12 (3): 463–476.

Chandler, D. (1999) *Bosnia: Faking Democracy after Dayton*. London: Pluto Press.

Chatterjee, P. (2004) *The Politics of the Governed: Reflections on Popular Politics in Most of the World (Leonard Hastings Schoff Lectures)*. New York: Columbia University Press.

Chivvis, C.S. (2010) 'Back to the Brink in Bosnia?', *Survival* 52 (1): 97–110.

Clarke, H. (2004) 'Privatization in Brcko District: Why it is different and why it works.' *Occasional Paper* 72, East European Studies at the Woodrow Wilson International Center, 20 pp.

Clinton, W. (1993) First Inaugural Address, Wednesday, January 21, 1993. http://www.bartleby.com/124/pres64.html (accessed 13 October 2011).

Cohen, J. and Arato, A. (1995) *Civil Society and Political Theory*. Cambridge, MA: MIT Press.

Cohen, L.J. (1995) *Broken Bonds: Yugoslavia's Disintegration and Balkan Politics in Transition*. Oxford: Westview Press.

Coleman, J. (1968) *Education and Politics*. Princeton: Princeton University Press.

Čolović, I. (2002) *The Politics of Symbol in Serbia*. London: Hurst & Co.

Commission of the European Communities (2003) Report from the Commission to the Council on the preparedness of Bosnia and Herzegovina to negotiate a

Stabilisation and Association Agreement with the European Union. Brussels: Commission of the European Communities, 46 pp.

Commission of the European Communities (2004) Communication from the Commission to the Council and the European Parliament: Area of Freedom, Security and Justice: Assessment of the Tampere programme and future orientations. Brussels: Commission of the European Communities, 17 pp.

Commission of the European Communities (2005) Communication from the Commission to the Council on the progress achieved by Bosnia and Herzegovina in implementing the priorities identified in the feasibility study on the preparedness of Bosnia and Herzegovina to negotiate a Stabilisation and Association Agreement with the European Union. Brussels: Commission of the European Communities, 6 pp.

Coppieters, B., Emerson, M., Huysseune, M., Kovziridze, T., Noutcheva, G., Tocci, N. and Vahl, M. (2004) *Europeanization and Conflict Resolution: Case Studies from the European Periphery*. Ghent: Academia Press.

Corbridge, S., Williams, G.G., Srivastava, M., and Véron, R. (2005) *Seeing the State: Governance and Governmentality in India*. Cambridge: Cambridge University Press.

Cornwall, A. and Brock, K. (2005) 'What do buzzwords do for development policy? A critical look at "participation", "empowerment" and "poverty reduction"', *Third World Quarterly* 26 (7): 1043–1060.

Council of Europe (2001) Bosnia and Herzegovina's application for membership of the Council of Europe. Strasbourg: Parliamentary Assembly of the Council of Europe, Document 9287 5 December 2001 http://assembly.coe.int/documents/workingdocs/doc01/edoc9287.htm (accessed 18 June 2012).

Coward, M. (2002) 'Community as heterogeneous ensemble: Mostar and multiculturalism', *Alternatives* 27: 29–66.

Coward, M. (2004) 'Urbicide in Bosnia', in S. Graham (ed.) *Cities, War and Terrorism*. Oxford: Blackwell, 154–171.

Cox, M. (2003) 'Building democracy from the outside: the Dayton Agreement in Bosnia and Herzegovina', in S. Bastian and R. Luckham (eds) *Can Democracy Be Designed? The Politics of Institutional Choice in Conflict-Torn Societies*. London: Zed Books, 253–276.

Crang, P. (1994) 'It's showtime: on the workplace geographies of display in a restaurant in southeast England', *Environment and Planning D: Society and Space* 12 (6): 675–704.

Cresswell, T. (1996) *In Place/Out of Place: Geography, Ideology, and Transgression*. Minneapolis: University of Minnesota Press.

Cresswell, T. (2002) 'Bourdieu's geographies: in memorium', *Environment and Planning D: Society and Space* 20 (3): 379–382.

Crook, R. (2001) 'Strengthening democratic governance in conflict torn societies: civic organisations, democratic effectiveness and political conflict', *IDS Working Paper* 129, 17 pp.

Dahlman, C. and Ó Tuathail, G. (2005a) 'Broken Bosnia: the localized geopolitics of displacement and return in two Bosnian places', *Annals of the Association of American Geographers* 95 (3): 644–662.

Dahlman, C. and Ó Tuathail, G. (2005b) 'The legacy of ethnic cleansing: the international community and the returns process in post-Dayton Bosnia-Herzegovina', *Political Geography* 24 (5): 569–599.

Dahlman, C.T. and Ó Tuathail, G. (2006) 'Bosnia's third space? Nationalist separatism and international supervision in Bosnia's Brčko District', *Geopolitics* 11 (4): 651–675.

de Souza Briggs, X. (1998) 'Doing democracy up-close: culture, power, and communication in community building', *Journal of Planning Education and Research* 18 (1): 1–13.

de Tocqueville, A. (1835) *De la Démocratie en Amérique*. Paris: Gosselin.

Delaney, D. (2001) 'Introduction: globalization and law', in N. Blomley, D. Delaney and R.T. Ford (eds) *The Legal Geographies Reader: Law, Power and Space*. Oxford: Blackwell, 252–255.

Derrida, J. (1994) *Specters of Marx: The State of Debt, the Work of Mourning and the New International* (trans. P. Kamuf). New York: Routledge.

Dittmer, J. and Dodds, K. (2008) 'Popular geopolitics past and future: fandom, identities and audiences', *Geopolitics* 13 (3): 437–457.

Dixon, K. (2008) 'New Labour and devolution: radicalism or bricolage?', *Textes et Contextes* 1: 33–40.

Dodds, K. (2005) 'Licensed to stereotype: geopolitics, James Bond and the spectre of Balkanism', *Geopolitics* 8 (2): 125–156.

Dolhinow, R. (2005) 'Caught in the middle: the state, NGOs, and the limits to grassroots organizing along the US–Mexico Border', *Antipode* 37 (3): 558–580.

Donais, T. (2000) 'Division and democracy: Bosnia's post-Dayton elections', in M. Spencer (ed.) *The Lessons of Yugoslavia*. London: JAI Elsevier Science, 229–258.

Donais, T. (2003) 'The political economy of stalemate: organised crime, corruption and economic deformation in post-Dayton Bosnia', *Conflict, Security & Development* 3 (3): 359–382.

Dowler, L. and Sharp, J. (2001) 'A feminist geopolitics?', *Space and Polity* 5 (3): 165–176.

Drakulić, S. (1996) *Café Europa: Life After Communism*. London: Abacus.

Drakulić, S. (2004) *They Would Never Hurt a Fly*. London: Abacus.

Dvornik, F. (1962) *The Slavs in European History and Civilization*. New Brunswick: Rutgers University Press.

Dyker, D.A. (1990) *Yugoslavia: Socialism, Development and Debt*. London: Routledge.

Džumhur, J. (2011) Introductory speech of Jasminka Dzumhur, Ombudsperson of Bosnia and Herzegovina at the OSCE Supplementary Human Dimension Meeting on National Human Rights Institutions. http://www.osce.org/odihr/77152 (accessed 5 October 2011).

Elden, S. (2009) *Territory and Terror: The Spatial Extent of Sovereignty*. Minneapolis: University of Minnesota Press.

Elden, S. (forthcoming) *The Birth of Territory*. Chicago: University of Chicago Press.

Emmert, T.A. (1990) *Serbian Golgotha: Kosovo, 1389*. Boulder, CO: East European Monographs.

Enck, G.E. and Preston, J.D. (1988) 'Counterfeit intimacy: a dramaturgical analysis of an erotic performance', *Deviant Behavior* 9 (4): 369–381.

Engler, S. (2003) 'Modern times: religion, consecration and the state in Bourdieu', *Cultural Studies* 17 (3/4): 445–467.

European Forum (1999) Social democratic parties in Bosnia and Hercegovina. www.europeanforum.bot-consult.se/cup/bosnia/socdem.htm (accessed 10 August 2005).

European Forum (2011) President Republica Srpska agrees to drop disputed referendum. 16 May. http://www.europeanforum.net/news/1153/president_republica_srpska_agrees_to_drop_disputed_referendum (accessed 18 August 2011).

European Union Special Representative (2011) *EUSR BiH and the EU.* http://www. delbih.ec.europa.eu/Default.aspx?id=10&lang=EN (accessed 3 February 2012).

Fagan, A. (2005) 'Civil society in Bosnia ten years after Dayton', *International Peacekeeping* 12 (3): 406–419.

Fagan, A. (2010) *Europe's Balkan Dilemma: Paths to Civil Society or State-building?* London: I.B. Tauris.

Featherstone, D. (2008) *Resistance, Space and Political Identities: The Making of Counter-Global Networks.* Oxford: Wiley-Blackwell.

Feit, B. and Morfit, M. (2002) 'A neutral space: the Arizona Market as an engine of peace and a new economy in Bosnia and Herzegovina.' Prepared for the USAID DG Partners Conference (Development Alternatives, Bethesda, MD, December).

Ferguson, J. (2006) *Global Shadows: Africa in the Neoliberal World Order.* Durham, NC: Duke University Press.

Ferguson, J.G. and Gupta, A. (2002) 'Spatializing states: toward an ethnography of neoliberal governmentality', *American Ethnologist* 29 (4): 981–1002.

Fine, B. (2002) 'It ain't social, it ain't capital and it ain't Africa', *Studia Africana* 13: 18–33.

Fleming, K.E. (2003) 'Orientalism, the Balkans, and Balkan historiography', *American Historical Review* 105 (4): 1–14.

Foucault, M. (1980) *Power/Knowledge.* New York: Pantheon.

Foucault, M. (1991) 'Governmentality', in G. Burchell, C. Gordon and P. Miller (eds) *The Foucault Effect: Studies in Governmentality.* Chicago: Chicago University Press, 87–104.

Franceschet, A. (2000) 'Popular sovereignty or cosmopolitan democracy? Liberalism, Kant and international reform', *European Journal of International Relations* 6 (2): 277–302.

Frost, A. and Yarrow, R. (1989) *Improvisation in Drama.* New York: St Martin's Press.

Fukuyama, F. (1995) 'Social capital and the global economy', *Foreign Affairs* 74 (5): 89–103.

Garvey, G. (1971) *Constitutional Bricolage.* Princeton: Princeton University Press.

Gibbon, E. (1828) *The History of the Decline and Fall of the Roman Empire,* vol. 3. London: Jones and Company.

Giddens, A. (1984) *The Constitution of Society: Outline of the Theory of Structuration.* Berkeley: University of California Press.

Glenny, M. (1999) *The Balkans: Nationalism, War and the Great Powers.* London: Granta Books.

Goffman, E. (1959) *The Presentation of Self in Everyday Life.* New York: Doubleday.

Goldsworthy, V. (1998) *Inventing Ruritania: The Imperialism of the Imagination.* London: Yale University Press.

Gow, J. (2003) *The Serbian Project and its Adversaries.* London: Hurst and Co.

Gregson, N. and Rose, G. (2000) 'Taking Butler elsewhere: performativities, spatialities and subjectivities', *Environment and Planning D: Society and Space* 18 (4): 433–452.

Griffiths, H. (1998) The dynamics of multi-national intervention: Brcko under international supervision. MPhil diss., University of Amsterdam.

Habermas, J. (2001) 'Why Europe needs a constitution', *New Left Review* 11: 5–26.

Hajek, A. (1993) 'Serbia', in M. Th. Houtsma, A. J. Wensinck and T. W. Arnold (eds) *E. J. Brill's First Encyclopaedia of Islam, 1913–1936*. Leiden: E. J. Brill, 198–205.

Hann, C. (1996) 'Introduction: political society and civil anthropology', in C. Hann and E. Dunn (eds) *Civil Society: Challenging Western Models*. London: Routledge, 1–26.

Hansen, L. (2006) *Security as Practice: Discourse Analysis and the Bosnian War*. London: Routledge.

Haraway, D. (1988) 'Situated knowledges: the science question in feminism and the privilege of partial perspective', *Feminist Studies* 14 (3): 575–599.

Harker, R. (1990) 'Bourdieu – education and reproduction', in R.M. Harker, C. Mahar and C. Wilkes (eds) *An Introduction to the Work of Pierre Bourdieu*. London: Macmillan.

Harriss, J. (2002) *Depoliticizing Development: The World Bank and Social Capital*. London: Anthem Press.

Harvey, D. (1976) 'A Marxian theory of the state', *Antipode* 8 (2): 80–89.

Hayden, R.M. (2002) 'Intolerant sovereignties and "multi-multi" protectorates: competition over religious sites and (in)tolerance in the Balkans', in C. Hann (ed.) *Postsocialism: Ideals, Ideologies and Practices in Eurasia*. London: Routledge, 159–179.

Hazan, P. (2004) *Justice in a Time of War: The True Story Behind the International Criminal Tribunal for the Former Yugoslavia*. College Station: Texas A&M University Press.

Hebdige, D. (1979) *Subculture: The Meaning of Style*. London: Methuen.

Held, D. and Archibugi, D. (1995) *Cosmopolitan Democracy: An Agenda for a New World Order*. Cambridge: Polity.

Helms, E. (2008) 'East and West kiss: gender, orientalism, and Balkanism in Muslim-majority Bosnia-Herzegovina', *Slavic Review* 67 (1): 88–119.

Herbert, S. (2000) 'For ethnography', *Progress in Human Geography* 24 (4): 550–568.

Hoare, M.A. (2010) 'The war of Yugoslav succession', in S.P. Ramet (ed.) *Central and Southeast European Politics Since 1989*. Cambridge: Cambridge University Press, 111–136.

Hobsbawm, E. and Ranger, T. (1983) 'Introduction: inventing traditions', in E. Hobsbawm and T. Ranger (eds) *The Invention of Tradition*. Cambridge: Cambridge University Press, 1–14.

Hodžić, R. (2010) 'Living the legacy of mass atrocities: victims' perspectives on war crimes trials', *Journal of International Criminal Justice* 8: 113–136.

Holbrooke, R. (1999) *To End A War*. New York: Modern Library.

Holmes, D. (2000) *Integral Europe: Fast-Capitalism, Multiculturalism and Neofascism*. Princeton: Princeton University Press.

Holston, J. (2008) *Insurgent Citizenship: Disjunctions of Democracy and Modernity in Brazil*. Princeton: Princeton University Press.

Human Rights Watch. (1992) *War Crimes in Bosnia-Hercegovina*. New York: Human Rights Watch.

Human Rights Watch (1996) *A Failure in the Making: Human Rights and the Dayton Agreement*. Sarajevo: Human Rights Watch.

Huntington, S.P. (1997) *The Clash of Civilizations and the Remaking of World Order*. New York: Simon & Schuster.

Husejnović, M. (2009) 'Awaiting state prison, Bosnia's Entity jails overflow', *Balkan Investigative Reporting Network*. 28 July. http://www.bim.ba/en/177/10/21363/ (accessed 20 August 2011).

Hyndman, J. (2003) 'Beyond either/or: a feminist analysis of September 11th', *Acme* 2: 1–13.

ICG (1996) Elections in Bosnia and Herzegovina – Part 2 of 3: The Lead-Up to Elections. Sarajevo: International Crisis Group, 25 pp.

ICG (1997) Going Nowhere Fast: Refugees and Internally Displaced Persons in Bosnia and Herzegovina. ICG Bosnia Report no. 23. 1 May. But it's 3+V+67 Sarajevo: International Crisis Group, 45 pp.

ICG (1998) Brcko: What Bosnia Could Be. ICG Bosnia Project – Report no. 31, 22 pp. Sarajevo: International Crisis Group, 1–21.

ICG (2003) Bosnia's Brcko: Getting in, Getting on and Getting out. Sarajevo/Brussels: International Crisis Group, 44 pp.

ICG (2009) Bosnia's Dual Crisis. Sarajevo/Brussels: International Crisis Group, 20 pp.

ICG (2011) Bosnia: State Institutions under Attack. http://www.crisisgroup.org/en/regions/europe/balkans/bosnia-herzegovina/b062-bosnia-state-institutions-under-attack.aspx (accessed 18 June 2012).

ICTY (1998) Drazen Erdemovic sentenced to 5 years of imprisonment. http://www.icty.org/sid/7686 (accessed 17 October 2011).

ICTY (2005a) Address of the Prosecutor at the Inauguration of the War Crimes Chamber of the Court of BH. http://www.icty.org/sid/8633 (accessed 17 October 2011).

ICTY (2005b) Trial Session 4th March 2005 Radovan Stanković. http://www.icty.org/x/cases/stankovic/trans/en/050304SC.htm (accessed 18 August 2011).

ICTY (2005c) Trial Session 15th March 2005 Slobodan Milošević. http://ictytranscripts.dyndns.org/trials/slobodan_milosevic/050315IT.htm (accessed 18 June 2012).

ICTY (2005d) Trial Session Case number 8 IT-00-39-T, the Prosecutor versus Momčilo Krajišnik. http://www.icty.org/x/cases/krajisnik/trans/en/050708ED.htm (accessed 17 August 2011).

ICTY (2006) President Tribunal President Welcomes Support for Bosnia And Herzegovina's Judicial Institutions. http://www.icty.org/sid/8776/en (Accessed 18 June 2012).

ICTY (2007) Trial Session for Dragomir Milošević 18th June 2007. http://www.icty.org/x/cases/dragomir_milosevic/trans/en/070618ED.htm (accessed 18 August 2011).

ICTY (2010) Mandate and Jurisdiction. http://www.icty.org/sid/320 (accessed 5 March 2010).

ICTY. (2012) Mandate and Jurisdiction. http://www.icty.org/sid/320 (accessed 10 May 2012).

Ingrao, C. (2011) 'Confronting the Yugoslav controversies: the Scholars' Initiative', in D. Šimko and U. Mäder (eds) *Stabilization and Progress in the Western Balkans*. Bern: Peter Lang, 139–162.

Ingrao, C. and T.A. Emmert (eds) (2009) *Confronting the Yugoslav Controversies: A Scholars' Initiative*. West Lafayette, IN: Purdue University Press.

Ivanić, M. (2005) 'The international community and Bosnia-Herzegovina', *Cambridge Review of International Affairs* 18 (2): 275–282.

Jambrek, P. (1975) *Development and Social Change in Yugoslavia: Crises and Perspectives of Building a Nation*. Toronto: Saxon House.

Jansen, S. (2007) 'Remembering with a difference: clashing memories of the Bosnian conflict in everyday life', in X. Bougarel, E. Helms and G. Duijzings (eds) *The*

New Bosnian Mosaic: Identities, Memories and Moral Claims in a Post-War Society. Aldershot: Ashgate, 193–210.

Jašarević, L. (2007) 'Everyday work: subsistence economy, social belonging and moralities of exchange at a Bosnian (black) market', in X. Bougarel, E. Helms and G. Duijzings (eds) *The New Bosnian Mosaic: Identities, Memories and Moral Claims in a Post-War Society.* Aldershot: Ashgate, 273–293.

Jeffrey, A. (2007) 'The politics of "democratization": lessons from Bosnia and Iraq', *Review of International Political Economy* 14 (3): 444–466.

Jeffrey, A. (2009a) 'Containers of fate: labelling states in the "war on terror"', in K. Dodds and A. Ingram (eds) *Spaces of Security and Insecurity: Geographies of the War on Terror.* Aldershot, Ashgate, 43–64.

Jeffrey, A. (2009b) 'Justice incomplete: Radovan Karadžić, the ICTY, and the spaces of international law', *Environment and Planning D: Society and Space* 27 (3): 387–402.

Jeffrey, A. (2011) 'The political geographies of transitional justice', *Transactions of the Institute of British Geographers* 36 (3): 344–359.

Jeffrey, A. and Jakala, M. (forthcoming) 'Beyond trial justice in the former Yugoslavia', *Geographical Journal.*

Jeffrey, C. (2001) '"A fist is stronger than five fingers": caste and dominance in rural north India', *Transactions of the Institute of British Geographers* 26 (2): 217–236.

Jeffrey, C. (2010) *Timepass: Youth, Class, and the Politics of Waiting in India.* Stanford: Stanford University Press.

Jeffrey, C., Jeffery, P. and Jeffery, R. (2008) *Degrees Without Freedom?.* Stanford: Stanford University Press.

Jessop, B. (1990) *State Theory: Putting the Capitalist State in its Place.* University Park, PA: Penn State University Press.

Jessop, B. (2001) 'Bringing the state back in (yet again): reviews, revisions, rejections, and redirections', *International Review of Sociology: Revue Internationale de Sociologie* 11 (2): 149–153.

Jessop, B. (2005) 'Critical realism and the strategic-relational approach', *New Formations* 56: 40–53.

Jessop, B. (2006) 'From micro-powers to governmentality: Foucault's work on statehood, state formation, statecraft and state power', *Political Geography* 26 (1): 34–40.

Jessop, B. (2008) *State Power.* Cambridge: Polity.

Jessop, B., Brenner, N. and Jones, M. (2008) 'Theorizing socio-spatial relations', *Environment and Planning D: Society and Space* 26 (3): 389–401.

Johnstone, D. (2002) *Fool's Crusade: Yugoslavia, NATO and Western Delusions.* London: Pluto Press.

Johnstone, K. (1981) *Impro: Improvisation and the Theatre.* London: Methuen Drama.

Jones, B. (forthcoming) '"Who does this District belong to?" Contesting, negotiating and practicing citizenship in a *mjesna zajednica* in Brčko District', *Transitions* 51 (1–2): 171–191.

Jones, R. (2007) *People/States/Territories.* Oxford: Wiley-Blackwell.

Jović, D. (2009) *Yugoslavia: A State that Withered Away.* West Lafayette, IN: Purdue University Press.

Judah, T. (2000) *The Serbs: History, Myth and the Destruction of Yugoslavia.* New Haven, CT: Yale University Press.

Juncos, A.E. (2005) 'The EU's post-conflict intervention in Bosnia and Herzegovina: (re)integrating the Balkans and/or (re)inventing the EU?', *Southeast European Politics* 6 (2): 88–108.

Kadrić, J. (1998) *Brčko: Genocide and Testimony*. Sarajevo: Institut for the Research of Crimes Against Humanity and International Law.

Kalajić, D. (1995) 'Srbi brane Evropu', *Duga*, 15 June.

Kalinić, D. (2004) Farewell address after being removed by Paddy Ashdown and US Ambassador Bond. Serbian Unity Congress. http://news.serbianunity.net/bydate/2004/July_02/6.html (accessed 20 August 2011).

Kaldor, M. (2003) *Global Civil Society: An Answer to War*. Cambridge: Polity Press.

Kaldor, M. and Vejvoda, I. (1997) 'Democratization in Central and Eastern European countries', *International Affairs* 73 (1): 59–82.

Kaletović, B. (2011) 'Political deadlock continues in BiH', *Southeast European Times* 3/12/11 http://setimes.com/cocoon/setimes/xhtml/en_GB/features/setimes/features/2011/12/03/feature-02 (accessed 18 June 2012).

Kaplan, R.D. (2005) *Balkan Ghosts: A Journey Through History*. New York: Vintage.

Karadžić, R. (1991) Work with a smile on your face. http://www.barnsdle.demon.co.uk/bosnia/kara.html (accessed 15 July 2011).

Kearns, G. (2009) *Geopolitics and Empire: The Legacy of Halford Mackinder*. Oxford: Oxford University Press.

Kermeli, E. (2008) 'Central administration versus provincial arbitrary governance: Patmos and Mount Athos monasteries in the 16th century', *Byzantine and Modern Greek Studies* 32 (2): 189–202.

Klemenčić, M. and Schofield, C. (1996) 'Mostar: make or break for the federation?', *Boundary and Security Bulletin* 4 (2): 72–76.

Krieger, H. (2001) *The Kosovo Conflict and International Law: An Analytical Documentation 1974–1999*. Cambridge: Cambridge University Press.

Kumar, R. (1997) *Divide and Fall? Bosnia in the Annals of Partition*. London: Verso.

Kundera, M. (1984) 'The tragedy of Central Europe', *New York Review of Books*, 26 April: 33–38.

Kuus, M. (2007) *Geopolitics Reframed: Security and Identity in Europe's Eastern Enlargement*. New York: Palgrave Macmillan.

Kuus, M. (2011) 'Bureaucracy and place: expertise in the European quarter', *Global Networks: Journal of Transnational Affairs* 11 (4): 421–439.

Lambourne, W. (2012) 'Outreach, Inreach and Civil Society Participation in Transitional Justice', in N. Palmer, P. Clark and D. Granville (eds) *Critical Perspectives in Transitional Justice*. Antwerp/Oxford: Intersentia, 235–262.

Lampe, J.R. (1996) *Yugoslavia as History: Twice There Was a Country*. Cambridge: Cambridge University Press.

Lévi-Strauss, C. (1972) *The Savage Mind*. London: Weidenfeld and Nicolson.

Lloyd, M. (1999) 'Performativity, parody, politics', *Theory, Culture & Society* 16 (2): 195–213.

Locke, G. (2002) *The Serbian Epic Ballads: An Anthology*. London: Association of Serbian Writers Abroad.

Lydall, H. (1984) *Yugoslav Socialism: Theory and Practice*. Oxford: Clarendon Press.

Mackenzie, D. (1996) *Violent Solutions: Revolutions, Nationalism, and Secret Societies in Europe to 1918*. Lanham, MD: University Press of America.

MacLean, F. (1949) *Eastern Approaches*. London: Jonathan Cape.

Mahar, C., Wilkes, C. and Harker, R. (1990) 'The basic theoretical position', in R.M. Harker, C. Mahar and C. Wilkes (eds) *An Introduction to the Work of Pierre Bourdieu*. London: Macmillan, 1–25.

Major, J. (1994) Mr Major's speech at Lord Mayor's Banquet. http://www.johnmajor.co.uk/page1380.html (accessed 14 October 2011).

Major, J. (1999) *John Major: The Autobiography*. London: Harper Collins.

Malcolm, N. (1994) *Bosnia: A Short History*. London: Macmillan.

Marston, S. (2004) 'Space, culture, state: uneven developments in political geography', *Political Geography* 23 (1): 1–16.

Marston, S.A. (1989) 'Public rituals and community power: St. Patrick's Day parades in Lowell, Massachusetts, 1841–1874', *Political Geography Quarterly* 8 (3): 255–269.

Matthias, J. and Vučković, V. (1987) *The Battle of Kosovo*. Athens, OH: Ohio University Press.

Mawdsley, E., Townsend, J.G., Porter, G., and Oakley, P. (2002) *Knowledge, Power and Development Agendas: NGOs North and South*. Oxford: INTRAC.

McConnell, F. (2009) 'De facto, displaced, tacit: the sovereign articulations of the Tibetan government-in-exile', *Political Geography* 28 (6): 343–352.

McConnell, F., Moreau, T. and Dittmer, J. (forthcoming) 'Mimicking state diplomacy: the legitimizing strategies of unofficial diplomacies', *Geoforum*.

McEvoy, K. (2007) 'Beyond legalism: towards a thicker understanding of transitional justice', *Journal of Law and Society* 34 (4): 411–440.

McFarlane, B. (1988) *Yugoslavia: Politics, Economics and Society*. London: Pinter.

McFarlane, C. (2009) 'Translocal assemblages: space, power and social movements', *Geoforum* 40 (4): 561–567.

McIlwaine, C. (1998) 'Civil society and development geography', *Progress in Human Geography* 22 (3): 415–424.

Megoran, N. (2006) 'For ethnography in political geography: experiencing and re-imagining Ferghana Valley boundary closures', *Political Geography* 25 (6): 622–640.

Mercer, C. (2002) 'NGOs, civil society and democratization: a critical review of the literature', *Progress in Development Studies* 2 (1): 5–22.

Mertus, J. (1999) *Kosovo: How Myths and Truths Started a War*. Berkeley: University of California Press.

Milošević, S. (1989) Milošević's Speech, 28 June 1989. http://emperors-clothes.com/milo/milosaid.html (accessed 17 September 2011).

Mitchell, T. (1991) 'The limits of the state – beyond statist approaches and their critics', *American Political Science Review* 85 (1): 77–96.

Mitchell, T. (1999) 'Society, economy, and the state effect', in G. Steinmetz (ed.) *State/Culture: State-Formation After the Cultural Turn*. Ithaca, NY: Cornell University Press, 76–97.

Mitchell, T. (2006) 'Society, economy and the state effect', in A. Sharma and A. Gupta (eds) *The Anthropology of the State: A Reader*. Oxford: Blackwell, 169–186.

Močnik, R. (2005) 'The Balkans as an element in ideological mechanisms', in D.I. Bjelic and O. Savic (eds) *Balkan as Metaphor Between Globalisation and Fragmentation*. Cambridge, MA: MIT Press.

Morris, J. (2002) *Trieste and the Meaning of Nowhere*. London: Faber and Faber.

Morrison, J. (1996) 'Bosnia's Muslim-Croat federation: unsteady bridge into the future', *Mediterranean Quarterly* 7 (1): 132–150.

Mouffe, C. (2005) *The Return of the Political*. London: Verso.

Mujkić, A. (2007) The Third Entity. http://www.pulsdemokratije.net/index. php?id=571&l=en (accessed 30 September 2009).

Müller, M. (2008) 'Reconsidering the concept of discourse for the field of critical geopolitics: towards discourse as language *and* practice', *Political Geography* 27: 322–338.

Murray Li, T. (1999) 'Compromising power: development, culture and rule in Indonesia', *Cultural Anthropology* 14 (3): 295–322.

Murray Li, T. (2007) *The Will to Improve: Governmentality, Development and the Practice of Politics*. Durham, NC: Duke University Press.

Nash, C. (2000) 'Performativity in practice: some recent work in cultural geography', *Progress in Human Geography* 24: 653–664.

Nava, M. (2002) 'Cosmopolitan modernity – everyday imaginaries and the register of difference', *Theory, Culture & Society* 19 (1–2): 81–100.

Navaro-Yashin, Y. (2002) *Faces of the State: Secularism and Public Life in Turkey*. Princeton: Princeton University Press.

Nayak, A. and Kehily, M.J. (2006) 'Gender undone: subversion, regulation and embodiment in the work of Judith Butler', *British Journal of Sociology of Education* 27: 459–472.

Nelson, L. (1999) 'Bodies (and spaces) do matter: the limits of performativity', *Gender, Place & Culture* 6 (4): 331–353.

Nezavisne novine. (2011) 'Nije uložen veto, slijedi referendum'. http://www.nezavisne.com/novosti/bih/Nije-ulozen-veto-slijedi-referendum-87032.html (accessed 9 September 2011).

Nolan, C.J. (2006) *The Age of Wars of Religions, 1000–1650: An Encyclopedia of Global Warfare and Civilization*. Westport, CT: Greenwood Press.

Nordstrom, C. (2003) 'Shadows and sovereigns', in N. Brenner, B. Jessop, M. Jones and G. MacLeod (eds) *State/Space: A Reader*. Oxford: Blackwell, 326–343.

Nussbaum, M. (1999) 'Professor of parody', *New Republic*, 22 February.

Ó Tuathail, G. (1986) 'The language and the nature of the "new" geopolitics: the case of US–El Salvador relations', *Political Geography Quarterly* 5: 73–85.

Ó Tuathail, G. (1994) 'The critical reading/writing of geopolitics: re-reading/writing Wittfogel, Bowman and Lacoste', *Progress in Human Geography* 18 (3): 313–332.

Ó Tuathail, G. (1996) *Critical Geopolitics*. London: Routledge.

Ó Tuathail, G. (2002) 'Theorizing practical geopolitical reasoning: the case of the United States' response to the war in Bosnia', *Political Geography* 21: 601–628.

Ó Tuathail, G. (2005) 'Embedding Bosnia-Herzegovina in Euro-Atlantic structures: from Dayton to Brussels', *Eurasian Geography and Economics* 46 (1): 51–67.

Ó Tuathail, G. and Agnew, J. (1992) 'Geopolitics and discourse: practical geopolitical reasoning in American foreign policy', *Political Geography* 11: 190–204.

Ó Tuathail, G., Dalby, S. and Routledge, P. (2006) *The Geopolitics Reader*, 2nd edn. Abingdon: Routledge.

Offe, C. (2004) 'Capitalism by democratic design? Democratic theory facing the triple transition in East Central Europe', *Social Research: An International Quarterly* 71 (3): 501–528.

Office of the High Representative (OHR) (1998) After the failure of his predecessors, Westendorp puts the Bosnian unity on the right path. Sarajevo: Office of the High Representative. http://www.ohr.int/ohr-dept/presso/pressa/default.asp?content_id=3174 (accessed 18 June 2012).

OHR (2000) *Bosnia and Herzegovina: Essential Texts*. Sarajevo: Office of the High Representative.

OHR (2001) *Essential Legal Texts*. Brčko: Office of the High Representative.

OHR (2002) Speech by the High Representative, Wolfgang Petritsch to the Council of Europe Parliamentary Assembly Concluding the debate on the accession of BiH to the Council. Sarajevo: Office of the High Representative.

OHR (2003) Supervisory Order on Additional Floors on Residential and Other Public Buildings. www.ohr.int (accessed 17 June 2011).

OHR (2005) Report to the European Parliament by the OHR and EU Special Representative for BiH, January–June 2005. Sarajevo: Office of the High Representative.

OHR (2006) Remarks by the Senior Deputy High Representative, Ambassador Peter Bas-Backer to the members of the NGO Council. http://www.ohr.int/ohr-dept/presso/presssp/default.asp?content_id=37296 (accessed 20 February 2011).

OHR (2010) RS Government Special Session: A Distasteful Attempt to Question Genocide. http://www.ohr.int/ohr-dept/presso/pressr/default.asp?content_id=44835 (accessed 20 August 2011).

Organization for Security and Cooperation in Europe (2011) Delivering Justice in Bosnia and Herzegovina. Sarajevo: Organization for Security and Cooperation in Europe, 1–105.

Owen, D. (1998) *Five Wars in the Former Yugoslavia, 1991–98*. Abu Dhabi: Emirates Center for Strategic Studies and Research, 28 pp.

Padgett, S. (1999) *Organizing Democracy in Eastern Germany: Interest Groups in Post-Communist Society*. Cambridge: Cambridge University Press.

Painter, J. (2000) 'Pierre Bourdieu', in M. Crang and N. Thrift (eds) *Thinking Space*. London: Routledge, 239–256.

Painter, J. (2002) 'Governmentality and regional economic strategies', in J. Hillier and E. Rooksby (eds) *Habitus: A Sense of Place*. Aldershot: Ashgate, 115–139.

Painter, J. (2003) 'State: society', in P. Cloke and R. Johnston (eds) *Spaces of Geographical Thought: Deconstructing Geography's Binaries*. London: Sage, 42–60.

Painter, J. (2006) 'Prosaic geographies of stateness', *Political Geography* 25 (7): 752–774.

Papandreou, M. (2000) 'Comparing the importance of international and domestic factors', in M. Spencer (ed.) *The Lessons of Yugoslavia: Research on Russia and Eastern Europe*, vol. 3. London: JAI Elsevier Science, 161–165.

Planning and Development Collaborative International (1988) PADCO: Planning and Development Collaborative International. http://pdf.usaid.gov/pdf_docs/PNABS678.pdf (accessed 17 October 2011).

Planning and Development Collaborative International (2005) Northeast Bosnia Local Government Support Activity: Final Report and Annual Report for Year 3. http://pdf.usaid.gov/pdf_docs/PDACG086.pdf (accessed 19 October 2011).

Pomerance, M. (1998) 'The Badinter Commission: the use and misuse of the International Court of Justice's jurisprudence', *Michigan Journal of International Law* 20: 50–57.

Porteous, J.D. and Smith, S. (2002) *Domicide: The Global Destruction of Home*. Montreal: McGill–Queen's University Press.

Postone, M., LiPuma, E. and Calhoun, C. (1993) 'Introduction: Bourdieu and social theory', in C. Calhoun, E. LiPuma and M. Postone (eds) *Bourdieu: Critical Perspectives*. Chicago: University of Chicago Press, 1–13.

Poulantzas, N. (1978) *State, Power, Socialism*. London: Verso.

Prpa-Jovanović, B. (2000) 'The making of Yugoslavia: 1830–1945', in J. Udovički and J. Ridgeway (eds) *Burn This House: The Making and Unmaking of Yugoslavia*. Durham, NC: Duke University Press, 43–63.

Pusić, E. (1975) 'Intentions and realities: local government in Yugoslavia', *Public Administration* 53 (2): 133–152.

Putnam, P. (1993) *Making Democracy Work: Civic Traditions in Modern Italy*. Princeton: Princeton University Press.

Radan, P. (2000) 'Post-secession international borders: a critical analysis of the opinions of the Badinter Arbitration Commission', *Melbourne University Law Review* 24 (1): 50–76.

Radcliffe, S. (2001) 'Imagining the state as space: territoriality and the formation of the state in Ecuador', in T.B. Hansen and F. Stepputat (eds) *States of Imagination: Ethnographic Explorations of the Postcolonial State*. Durham, NC: Duke University Press, 122–145.

Radu, C. (2011) 'Governmentality and the deportation of Eastern European Roma in Italy and France', *Student Pulse* 3 (4), http://www.studentpulse.com/articles/513/2/governmentality-and-the-deportation-of-eastern-european-roma-in-italy-and-france (accessed 10 October 2011).

Raento, P. (2006) 'Communicating geopolitics through postage stamps: the case of Finland', *Geopolitics* 11 (4): 601–629.

Ramet, P. (1985) 'Apocalypse culture and social change in Yugoslavia', in P. Ramet (ed.) *Yugoslavia in the 1980s*. Boulder, CO: Westview Press.

Rawls, A.W. (1987) 'The interaction order sui generis: Goffman's contribution to social theory', *Sociological Theory* 5 (2): 136–149.

Research and Documentation Center Sarajevo (2007) Bosnian Book of the Dead. http://www.idc.org.ba/ (accessed 8 March 2012).

Robinson, B. (2004) 'Putting Bosnia in its place: critical geopolitics and the representation of Bosnia in the British print media', *Geopolitics* 9 (2): 378–401.

Robinson, G., Engelstift, S. and Pobrić, A. (2001) 'Remaking Sarajevo: Bosnian nationalism after the Dayton Accord', *Political Geography* 20 (8): 957–980.

Rose, G. (1997) 'Situating knowledges: positionality, reflexivities and other tactics', *Progress in Human Geography* 21 (3): 305–320.

Rose, G. (1998) 'The exit strategy delusion', *Foreign Affairs* 77 (1): 56–67.

Rosenbaum, A. (2001) 'Building equitable and effective local governance in a complex ethnic environment: the case of Brcko Government'. www.dec.org/pdf_docs/PDABY137.

Ross, C., Mirowsky, J. and Pribesh, S. (2001) 'Powerlessness and the amplification of threat: neighborhood disadvantage, disorder, and mistrust', *American Sociological Review* 66: 568–591.

Said, E. (1978) *Orientalism*. London: Penguin.

Scheper-Hughes, N. (1992) *Death Without Weeping: The Violence of Everyday Life in Brazil*. Berkeley: University of California Press.

Schwarz-Schilling, C. (2006) Bosnia's Way Forward, in *Internationale Politik* Spring 2006, available at https://ip-journal.dgap.org/en/ip-journal/regions/bosnias-way-forward.

Scott, J.C. (1985) *Weapons of the Weak: Everyday Forms of Peasant Resistance*. New Haven, CT: Yale University Press.

Scott, J.C. (1998) *Seeing Like a State: How Certain Schemes to Improve the Human Condition Have Failed*. New Haven, CT: Yale University Press.

Segal, L. (2008) 'After Judith Butler: identities, who needs them?', *Subjectivity* 25: 381–394.

Shapiro, M. and Hacker-Córden, C. (1999) *Democracy's Value*. Cambridge: Cambridge University Press.

Sharma, A. and Gupta, A. (2006) 'Introduction: rethinking theories of the state in an age of globalisation', in A. Sharma and A. Gupta (eds) *The Anthropology of the State: A Reader*. Oxford: Blackwell, 1–42.

Sherwell, P. (2000) 'Guns, girls, drugs, fake track suits: it's all here in the wildest market in the world', *Sunday Telegraph*, London, 19 November.

Shore, C. (2000) *Building Europe: The Cultural Politics of European Integration*. London: Routledge.

Silber, L. and Little, A. (1996) *The Death of Yugoslavia*. London: Penguin.

Simms, B. (2001) *Unfinest Hour: Britain and the Destruction of Bosnia*. London: Penguin.

Singleton, F. (1976) *Twentieth-Century Yugoslavia*. London: Macmillan.

Smith, N. (2003) *American Empire: Roosevelt's Geographer and the Prelude to Globalisation*. Berkeley: University of California Press.

Sokolović, D. (2005) 'How to conceptualize the tragedy of Bosnia: civil, ethnic, religious war or . . .?', *War Crimes, Genocide, & Crimes against Humanity* 11 (1): 115–130.

Sommers, W. (2002) *Brčko District: Experiment to Experience*, available at http://www.esiweb.org/pdf/bridges/bosnia/Sommers BrckoENG/pdf (accessed 18 June 2012).

Somogyi, L. (1993) 'Introduction', in L. Somogyi (ed.) *The Political Economy of the Transition Process in Eastern Europe*. Aldershot: Edward Elgar, 1–54.

Stability Pact (2006) *The Stability Pact for South Eastern Europe*. Brussels: Stability Pact.

Stankovic, M. (2000) *Trusted Mole*. London: HarperCollins.

Stoker, B. (2003[1847]) *Dracula*. New York: Barnes and Noble Classics.

Superbosna (2003) Spomen ploča Srbima donešena i odnešena. http://www.super bosna.com/vijesti/ostalo/spomen_plo%E8a_srbima_done%B9ena_i_ odne%B9ena/ (accessed 18 August 2011).

Taussig, M. (1997) *The Magic of the State*. London: Routledge.

Taylor, P. (1994) 'The state as container: territoriality in the modern state system', *Progress in Human Geography* 18 (2): 151–162.

Tepavać, M. (2000) 'Tito: 1945–1980', in J. Udovički and J. Ridgeway (eds) *Burn This House: The Making and Unmaking of Yugoslavia*. Durham, NC: Duke University Press.

Thrift, N. (2003) 'Performance and . . .', *Environment and Planning A* 35: 2019–2024.

Thrift, N. and Dewsbury, J.D. (2000) 'Dead geographies – and how to make them live', *Environment and Planning D: Society and Space* 18 (4): 411–432.

Tito, J.B. (1974) 'Speech by Tito at the Tenth Congress of the League of Communist Yugoslavia', *Socialist Thought and Practice* 6–7: 45–47.

Toal, G. and Dahlman, C. (2011) *Bosnia Remade: Ethnic Cleansing and Its Reversal*. New York: Oxford University Press.

Todorova, M. (1997) *Imagining the Balkans*. New York: Oxford University Press.

Trouillot, M.-R. (2001) 'The anthropology of the state in the age of globalization', *Current Anthropology* 42 (1): 125–138.

Tvedt, T. (2002) 'Development NGOs: actors in global civil society or in a new international social system?', *Voluntas: International Journal of Voluntary and Nonprofit Organisations* 13 (4): 363–375.

Udovički, J. and J. Ridgeway (eds) (2000) *Burn This House: The Making and Unmaking of Yugoslavia*. Durham, NC: Duke University Press.

Udovički, J. and Stikovać, E. (2000) 'Bosnia and Hercegovina: the second war', in J. Udovički and J. Ridgeway (eds) *Burn This House: The Making and Unmaking of Yugoslavia*. Durham, NC: Duke University Press, 175–216.

UNDP (2001) Interim Report. Brčko Local Action Programme (BLAP). Sarajevo: UNDP, 75 pp.

UNDP (2002) *Human Development Report 2002: Bosnia and Herzegovina*. Sarajevo: Economics Institute Sarajevo, 113 pp.

UNDP (2009) *The Ties that Bind: Social Capital in Bosnia and Herzegovina*. Sarajevo: UNDP.

UNHCR (1997a) Bosnia and Herzegovina, Repatriation and Return Operation. New York: United Nations High Commission for Refugees.

UNHCR (1997b) Reception Capacity/Target Areas – Working Document. New York: United Nations High Commission for Refugees.

UNHCR (2005) Return Statistics 1996–2005. www.unhcr.ba/returns/T4-0305.pdf.

UNICEF (2002) Trafficking in Human Beings in Southeastern Europe. Belgrade: UNICEF, UNOHCHR, OSCE-ODIHR.

United Nations Security Council (1993) Resolution 827. http://www.un.org/Docs/scres/1993/scres93.htm (accessed 18 February 2010).

United Nations Security Council (2003) Security Council Briefed on Establishment of War Crimes Chamber within State Court of Bosnia And Herzegovina. http://www.un.org/News/Press/docs/2003/sc7888.doc.htm (accessed 14 June 2011).

van Genugten, S. (2011) 'Libya after Gadhafi', *Survival* 53: 61–74.

Velek, J. (2004) 'Juergen Habermas and the utopia of perpetual peace', *Filosoficky Casopis* 52 (2): 231–256.

Voss, M. (2001) '"Slave trade" thrives in Bosnia'. http://news.bbc.co.uk/1/hi/world/europe/1209085.stm (accessed 16 June 2012).

Vukić, U. (2011) Pola RS za ukidanje Suda BiH. *Nezavisne novine*, 18 April.

Wachtel, A. and Bennett, C. (2009) 'The dissolution of Yugoslavia', in C. Ingrao and T.A. Emmert (eds) *Confronting the Yugoslav Controversies: A Scholar's Initiative*. West Lafayette, IN: Purdue University Press, 12–47.

Weber, C. (1998) 'Performative states', *Millennium: Journal of International Studies* 27: 77–95.

Weber, M. (1958) 'Politics as a vocation', in H.H. Gerth and C. Wright Mills (eds) *From Max Weber: Essays in Sociology*. New York: Oxford University Press, 77–128.

Weick, K.E. (1998) 'Introductory essay: improvisation as a mindset for organizational analysis', *Organization Science* 9 (5): 543–555.

Werbner, R. (1986) 'Review article: the political economy of bricolage', *Journal of Southern African Studies* 13 (1): 151–156.

West, R. (2001[1941]) *Black Lamb and Grey Falcon*. Edinburgh: Canongate.

White, H. (1973) *Metahistory: The Historical Imagination in Nineteenth-Century Europe*. Baltimore: Johns Hopkins University Press.

Williams, M. (1996) 'Polling for partition', *Index on Censorship* 25: 13–26.

Willke, H. (2010) 'Transparency after the financial crisis: democracy, transparency, and the veil of ignorance', in S.A. Jansen, E. Schröter and N. Stehr (eds) *Transparenz: Multidisciplinäre Durchsichten durch Phänomene und Theorien des Undurchsichtigen*. Jahrbuch der Zeppelin University Friedrichshafen. Wiesbaden: VS-Verlag, 56–81.

Wisler, D. (2007) 'The international civilian police mission in Bosnia and Herzegovina: from democratization to nation-building', *Police Practice and Research* 8 (3): 253–268.

Woodward, S. (1995) *Balkan Tragedy: Chaos and Dissolution after the Cold War*. Washington, DC: Brookings Institution.

Woodward, S. (1996) 'The West and international organisations', in D.A. Dyker and I. Vejvoda (eds) *Yugoslavia and After: A Study in Fragmentation, Despair and Rebirth*. London: Longman, 155–178.

Woodward, S. (1997) 'Bosnia after Dayton: year two', *Current History* 96 (608), 97–103.

World Bank (1996) *The Initiative on Defining, Monitoring and Measuring Social Capital*. Washington, DC: World Bank.

Yiftachel, O. (2009) 'Critical theory and "gray space": mobilization of the colonized', *City* 13 (2–3): 246–263.

Zimmerman, W. (1996) *Origins of a Catastrophe: Yugoslavia and Its Destroyers – America's Last Ambassador Tells What Happened and Why*. London: Times Books.

Žižek, S. (1990) 'Eastern Europe's Republic of Gilead', *New Left Review* 183: 50–62.

Zolo, D. (2002) *Invoking Humanity: War, Law, and Global Order*. London: Continuum.

Index

The Improvised State: Sovereignty, Performance and Agency in Dayton Bosnia, First Edition. Alex Jeffrey.
© 2013 John Wiley & Sons, Ltd. Published 2013 by John Wiley & Sons, Ltd.